Royal and Urban Gunpowder Weapons in Late Medieval England

Armour and Weapons
ISSN 1746-9449

Series Editors
Kelly DeVries
Robert W. Jones
Robert C. Woosnam-Savage

Throughout history armour and weapons have been not merely the preserve of the warrior in battles and warfare, but potent symbols in their own right (the sword of chivalry, the heraldic shield) representing the hunt and hall as well as the battlefield. This series aims to provide a forum for critical studies of all aspects of arms and armour and their technologies, from the end of the Roman Empire to the dawn of the modern world; both new research and works of synthesis are encouraged.

New proposals for the series are welcomed; they should be sent to the publisher at the address below.

Boydell & Brewer Limited, PO Box 9, Woodbridge, Suffolk, IP12 3DF

Also in this series:

The Artillery of the Dukes of Burgundy, 1363–1477,
Robert Douglas Smith and Kelly DeVries
'*The Furie of the Ordnance': Artillery in the English Civil Wars*, Stephen Bull
Jousting in Medieval and Renaissance Iberia, Noel Fallows
The Art of Swordsmanship *by Hans Lecküchner*, translated by Jeffrey L. Forgeng
The Book of Horsemanship *by Duarte I of Portugal*,
translated by Jeffrey L. Forgeng
Pietro Monte's Collectanea: *The Arms, Armour and Fighting Techniques of a Fifteenth-Century Soldier*, translated by Jeffrey L. Forgeng
The Medieval Military Engineer: From the Roman Empire to the Sixteenth Century,
Peter Purton

Royal and Urban Gunpowder Weapons in Late Medieval England

Dan Spencer

THE BOYDELL PRESS

First published 2019
The Boydell Press, Woodbridge

ISBN 978 1 78327 457 4

The Boydell Press is an imprint of Boydell & Brewer Ltd
PO Box 9, Woodbridge, Suffolk IP12 3DF, UK
and of Boydell & Brewer Inc.
668 Mt Hope Avenue, Rochester, NY 14620–2731, USA
website: www.boydellandbrewer.com

A CIP catalogue record for this book is available
from the British Library

This publication is printed on acid-free paper

MIX
Paper from
responsible sources
FSC® C013056

Printed and bound in Great Britain by TJ International Ltd, Padstow, Cornwall

To my mother, Elizabeth Spencer, for always being there for me

Contents

Illustrations

PLATES

MAPS

FIGURES

TABLES

The author and publisher are grateful to all the institutions and individuals listed for permission to reproduce the materials in which they hold copyright. Every effort has been made to trace the copyright holders; apologies are offered for any omission, and the publisher will be pleased to add any necessary acknowledgement in subsequent editions.

Acknowledgements

THIS BOOK OWES ITS existence to discussions I had with Professor Anne Curry in 2012, who suggested that I should investigate the early history of English gunpowder weapons. At the time I had little understanding of the sheer scale of the topic and how much the project would grow and expand over time. It has proved to be a particularly fruitful area of research, which has required many visits to numerous archives across England and countless hours spent transcribing, translating and analysing medieval texts. This book would not have been finished without the support and guidance of the people who have assisted me with this project. Firstly, I would like to thank the staff at Boydell & Brewer, especially Caroline Palmer, for their assistance in publishing this work. I am grateful to Dr Alexander Brondarbit, Dr Adam Chapman, Mr Nicholas Kingwell and Professor Emmanuel de Crouy-Chanel for contributing a number of archival references. Thanks should also go to Professor Peter Clarke and the other members of the Centre for Medieval and Renaissance Culture at the University of Southampton for offering valuable training in Medieval Latin and palaeography. Further feedback was provided by Professor Steven Gunn and Dr Craig Lambert to whom I am indebted. I am particularly appreciative of the support provided by Professor Anne Curry and Professor Mark Stoyle, for their invaluable advice, guidance and constructive criticism of my work. I am also grateful to my partner, Hannah, for her support and encouragement in completing this project. Finally, I would like to thank my mother, Elizabeth Spencer, for the time and effort that she spent in proofreading and commenting on drafts of this work. This book is dedicated to her for always having been there for me.

Abbreviations

BL	British Library
BNF	Bibliothèque Nationale de France
CCA	Canterbury Cathedral Archive
CCR	Calendar of Close Rolls
CPR	Calendar of Patent Rolls
DHC	Devon Heritage Centre
DUL	Durham University Library
ESRO	East Sussex Record Office
HHC	Hull History Centre
HRO	Hampshire Record Office
KHLC	Kent History and Library Centre
KLBA	King's Lynn Borough Archives
LMA	London Metropolitan Archives
MCO	Magdalen College Oxford
NRO	Norfolk Record Office
PWDRO	Plymouth and West Devon Record Office
SRO	Southampton Record Office
TNA	The National Archives
WALMR	Westminster Abbey Library and Muniment Room
WSRO	Wiltshire and Salisbury Record Office

A Note on Terms Used

A VARIETY OF WORDS were used to describe gunpowder weapons in the Middle Ages. This was due, at least in part, to the use of three languages for financial accounts in this period: English, Anglo-Norman and Latin. The chronicler Walsingham illustrated this point when he referred to 'gunnas … quas Galli "canones" vocant'.[1] In addition to 'gun' and 'cannon', the term 'artillery' (*artillaria* in Latin) was used to refer to all types of missile weapons including guns.[2] By the fifteenth century, the word ordnance (also of Latin origin) was employed when describing military equipment including firearms.[3] For the purposes of this discussion, the term gun will be used in preference to cannon, as the former word is English, whereas ordnance and artillery will be used to refer to gunpowder weapons unless otherwise stated. The names used to describe different types of guns have also been standardised to avoid confusion. A full discussion of the terms relating to ordnance can be found in the glossary with further information available in Appendix J. A variety of terms were also used to describe guns made of copper-alloys including copper (*cupra*), brass (*aenum/eneus*), laton (*latone*) and metal (*metallum*), with the modern word bronze being used for consistency.[4]

1 H. T. Riley, ed., *Thomae Walsingham, Quondam Monachi S. Albani, Historia Anglicana, Volume 2: A. D. 1381–1422* (Cambridge: Cambridge University Press, 2012), p. 323.

2 T. F. Tout, 'Firearms in England in the Fourteenth Century', *The English Historical Review*, 26 (1911), 666–702 (p. 679).

3 For the forms of these Latin words see R. E. Latham, *Revised Medieval Latin Word-List: From British and Irish Sources* (London: Oxford University Press, 1965), pp. 32, 324–5.

4 For a full discussion of copper-alloy terminology see Chapter 8.

Glossary

(*Further detailed information about the gun types marked with asterisks can be found in Appendix J)

Artillery A term referring to all types of missile weapons including guns.[1]

Artillery tower A masonry tower constructed specifically for the use of firearms, with provision for gunports or platforms. Examples of surviving artillery towers in England include the Cow Tower in Norwich and God's House Tower in Southampton.[2]

Bastard gun A non-standard type of gun.[3]

Bombard* The largest category of guns, also commonly referred to as large guns (see Plates 2 and 6).

Bombardelle* A smaller version of a bombard.

Boulevard French word for **Bulwark**.

Bronze A copper alloy which includes a small quantity of tin, which was generally referred to as brass in financial accounts.

Bulwark An artillery fortification designed for the deployment of guns. These were constructed out of wood, stones or bricks and were typically erected in front of gatehouses of towns or castles.

Cannon The French word for gun.

Carriage See **Cart**.

Cart A vehicle used for transporting guns drawn by horses or oxen. In the first half of the century guns were moved in large carts and were then attached to their

1 Tout, 'Firearms in England in the Fourteenth Century', *The English Historical Review*, 26 (1911), p. 679.

2 See Chapters 6 and 7.

3 Howard L. Blackmore, *The Armouries of the Tower of London, I: Ordnance* (London: H. M. S. O., 1976), p. 219.

wooden beds known as 'stocking'. Later in the century they were given their own individual horse drawn carriages.

Cart gun* A type of gun mounted on a cart. Initially described as a *ribaudequin*, or a variant of this term.

Charcoal A form of fuel derived from wood which serves as a component of gunpowder.

Charging ladle A ladle used for placing gunpowder into the gun.[4]

Crapaudeau A long type of gun with a small bore. This gun was employed by the French and Burgundians in the fifteenth century but not by the English.[5]

Culverin* A long gun with a small bore.

Curtow* A large gun which was used in the late fifteenth and early sixteenth centuries.

Demi-curtow* A gun half the size of a curtow.

Falcon* A small type of field gun dating from the late fifteenth century.

Founder A worker employed in the casting of metal.

Fowler* This was originally a small type of gun dating from the early fifteenth century. Later in the century new larger variants of this gun were developed. The French term for the gun was **Veuglaire**.

Gun The standard English word for a firearm.

Gun chamber A removable powder chamber.[6]

Gun-maker A professional worker who constructed guns.

Gunner A professional artisan employed in the construction, maintenance and operation of guns and gunpowder.

Gunport A form of defensive architecture, typically a hole carved in a stone wall for the use of firearms.

Gunpowder An incendiary mixture of saltpetre, sulphur and charcoal which was used to ignite guns. By the 1480s this was stored in a standard format, with a barrel containing 200lbs and a last containing 2,400lbs.[7] Also referred to as black powder.

Gunstone The standard form of ammunition for most guns in the fifteenth century.

Hackbut* A type of heavy handgun, that was supported by a hook and which had a tiller.

4 Blackmore, *The Armouries of the Tower of London*, p. 232.

5 Robert D. Smith and Kelly DeVries, *The Artillery of the Dukes of Burgundy 1363–1477* (Woodbridge: Boydell Press, 2005), p. 221.

6 Blackmore, *The Armouries of the Tower of London*, p. 223.

7 TNA, E 101/198/13, f. 6; E 36/8, f. 2v.

Handgun* A small type of firearm which could be operated and moved by hand. In the late fifteenth century this term was used to refer to an arquebus type of handgun.

Iron Dice Appears to be an early form of hail shot used to target enemy personnel at short range.[8]

Iron Shot A projectile made out of iron. Prior to the reign of Henry VII, small quantities of wrought iron shot were used; by the 1490s cast iron shot was cast in Ashdown Forest in East Sussex.

Large gun see **Bombard***.

Lead Shot The most common form of projectile for small guns and handguns.

Mantlet A large wooden structure created by carpenters which was used to protect guns during a siege.

Master of the Ordnance An official responsible for the administration of the ordnance.[9]

Miche A swivel mount for guns.[10]

Murderer* A large type of gun. (The term is also used to describe a type of artillery fortification.)

Ordnance A term used for military equipment including firearms.

Ordnance company A mixed company of different professionals employed to operate and maintain artillery, typically this included gunners, smiths, carpenters, masons and labourers.[11]

Organ guns* A type of gun which consisted of numerous small barrels.

Pavise A large wooden shield which could be fixed to the ground.

Pellet gun* A small gun that fired lead shot.

Pot gun* A large gun which was possibly used as a mortar.

Privy wardrobe An office responsible for the production, storage and maintenance of royal firearms from the late fourteenth century into the early fifteenth century.

Quarrel gun* An early type of gun that fired quarrels.

Rabbet 'To join or fix by means of a rabbet or rabbets.'[12]

8 Glenn Foard and Anne Curry, *Bosworth 1485: A Battlefield Rediscovered* (Oxford: Oxbow Books, 2013), p. 149.

9 Andy King, 'Gunners, Aides and Archers: The Personnel of the English Ordnance Companies in Normandy in the Fifteenth Century', *Journal of Medieval Military History*, 9 (2011), 65–75.

10 Blackmore, *The Armouries of the Tower of London*, p. 234.

11 King, 'Gunners, Aides and Archers'.

12 C. T. Onions, ed., *The Shorter Oxford English Dictionary*, volume 2 (Oxford: Clarendon Press, 1973), p. 1735.

Rammer A block of wood used to ram the charge of a gun down the barrel.[13]

Ribaudequin see **Cart gun***.

Saltpetre Potassium nitrate, a component of gunpowder.

Serpentine* A long thin type of gun first recorded in English accounts in 1450 (see Plate 1).

Smith A professional skilled in iron work, who could be employed in the forging of **Wrought iron** guns.

Springald A large torsion form of mechanical artillery designed for firing large bolts.[14]

Stock A wooden bed or support for a gun.[15]

Tampon A wad made out of wood, which was used to separate the gunpowder from the projectile.[16]

Trestle gun* A small type of gun placed on a trestle.

Veuglaire French word for **Fowler***.

Wrought iron A type of low quality malleable iron, which was the most common material for the construction of guns in the fifteenth century.

13 Blackmore, *The Armouries of the Tower of London*, p. 240.

14 Thom Richardson, *The Tower Armoury in the Fourteenth Century* (Leeds: Royal Armouries, 2016), pp. 133–40.

15 Blackmore, *The Armouries of the Tower of London*, p. 245.

16 Ibid., p. 246.

Plate 1 Artist depiction of a serpentine mounted on a wooden carriage.
Drawn by Scott Hall © Dan Spencer.

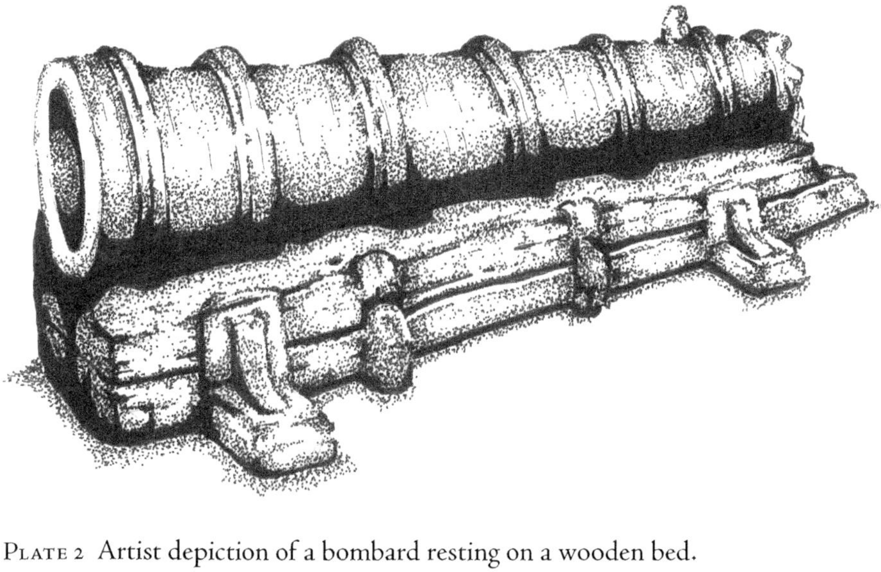

Plate 2 Artist depiction of a bombard resting on a wooden bed.
Drawn by Scott Hall © Dan Spencer.

PLATE 3 A gunport in the Arcades at Southampton. By permission of Southampton City Council.

PLATE 4 God's House Tower, Southampton. An early fifteenth-century artillery tower which was built for the deployment and storage of the town's ordnance. By permission of Southampton City Council.

Plate 5 A miniature depicting the use of mounted guns during a siege. William of Tyre, *Histoire d'Outremer or Historia rerum in partibus transmarinis gestarum* (c.1479–80). BL, Royal MS 15 E I, f. 357. Creative Commons CC0 1.0 Universal Public Domain Dedication.

PLATE 6 A miniature depicting the use of bombards resting on beds during a siege. Jean Froissart, *Chroniques*, volume 4, part 1 (the 'Harley Froissart') (c.1470–2). BL, Harley MS 4379, f. 83v. Creative Commons CC0 1.0 Universal Public Domain Dedication.

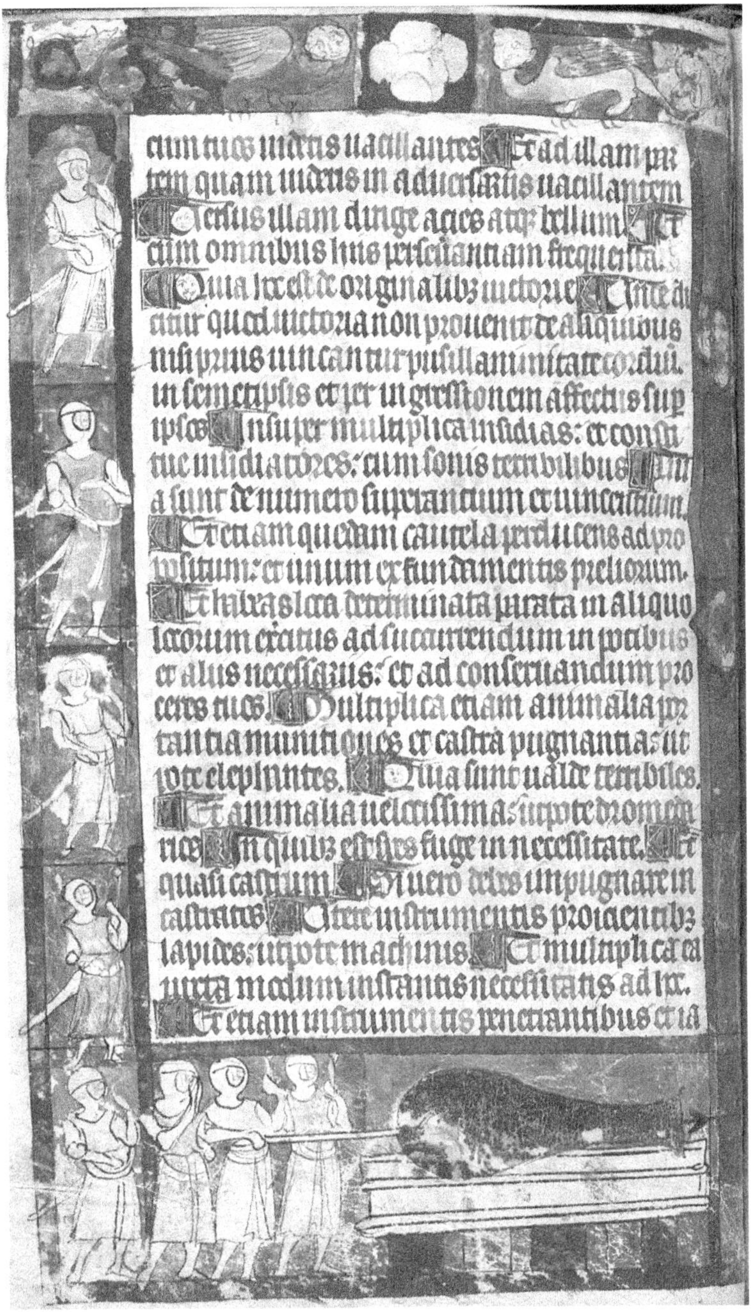

PLATE 7 A miniature depicting soldiers operating an early gun which fires arrow-shaped projectiles. Pseudo-Aristotle, *Secretum Secretorum* (c.1326–7). BL, Additional MS 47680, f. 44v. Creative Commons CC0 1.0 Universal Public Domain Dedication.

Introduction

One of the most significant developments of the Middle Ages was the adoption and use of gunpowder artillery. This technology originated in China and had reached Western Europe by the early fourteenth century.[1] One of the earliest known images of a European gun is depicted in the treatise *De Nobilitatibus, Sapientiis, et Prudentiis Regum* given by the scholar Walter de Milemete to Edward III on his accession to the throne in 1327. This shows a cauldron-shaped gun lying on a trestle which is firing a quarrel or arrow-shaped projectile at a building.[2] Over time the new technology gradually transformed the conduct of warfare in Western Europe, at first by supplanting traditional mechanical artillery and then personal missile weapons, such as the crossbow or longbow. Guns eventually came to be used in all aspects of military operations, in sieges, on the battlefield and in naval warfare, with kings, nobles and burgesses all spending large sums of money on these prestigious weapons. The growing effectiveness of gunpowder artillery also prompted significant changes in the conduct and organisation of warfare. New types of professionals were employed to manufacture, repair and operate guns, with the need to manage and organise artillery efficiently leading to the development of new administrative systems. This was necessary as the logistical requirements of transporting and supplying ordnance to armies, fleets and garrisons increased over time. In the case of transportation, the provision of large numbers of carts, pack animals and auxiliary workers became essential for the effective operation of field trains during campaigns. The fortifications of towns and castles were also adapted to better withstand gunshot and to maximise the use of defensive firepower, which led to radical changes in the design of these structures. These changes contributed to the increasing cost of waging war for late medieval states, whose rulers were

1 For a discussion of the Chinese use of gunpowder weapons see Joseph Needham, Ho Ping Yú, Lu Gwei-Djen and Wang Ling, *Science and Civilisation in China. Vol. 5, Chemistry and Chemical Technology*, part 7, 'Military Technology: The Gunpowder Epic' (Cambridge: Cambridge University Press, 1986).

2 For a similar depiction see Plate 7. Hugh Pollard, *Pollard's History of Firearms* (London: Country Life Books, 1983), p. 28.

obliged to adopt new measures to increase their incomes, to ensure that they could keep pace with their rivals.

The significance of gunpowder artillery has long been recognised by historians of the Middle Ages.[3] Surprisingly, despite this, the English experience of using firearms lacks its own specialist treatment in a monograph, although certain areas have been examined in some detail.[4] The early history of royal gunpowder weapons in England was discussed at length by Tout in an important article published in 1911.[5] Tout examined the valuable information contained within the accounts submitted by the keepers of the privy wardrobe, an organisation responsible for the management and procurement of royal weaponry, which he drew upon to trace the history of the technology throughout the fourteenth century.[6] This allowed him to demonstrate that the development of firearms in England was very similar to that of the Continent. Yet the lack of a comparable body of documentary evidence for centrally held royal gunpowder weapons in England in the fifteenth century has meant that later authors have tended to focus on studying specific aspects of the topic, such as the weaponry possessed by the Calais garrison.[7] This is demonstrated in works by David Grummitt, who made use of the extensive surviving documentation submitted by the treasurers and victuallers of Calais to argue that English artillery was comparable to that of her European rivals in the fifteenth century.[8]

3 For instance see Charles Oman, *A History of the Art of War in the Middle Ages, Volume Two: 1278–1485 AD* (London: Greenhill Books, 1998), pp. 226–9.

4 The topic is discussed by O. F. G. Hogg in two works, but his findings were based on very limited research and so are of little value. O. F. G. Hogg, *English Artillery 1326–1716* (London: Royal Artillery Institution, 1963); O. F. G. Hogg, *Artillery: Its Origin, Heyday and Decline* (London: C. Hurst, 1970).

5 T. F. Tout, 'Firearms in England in the Fourteenth Century', *The English Historical Review*, 26 (1911), 666–702. See also Joseph Hunter, 'Proofs of the Early Use of Gunpowder in the English Army', *Archaeologia*, 32/2 (1846), 379–87; H. Brackenbury, 'Ancient Cannon in Europe. Part I: From their First Employment to A.D. 1350', *Proceedings of the Royal Artillery Institution*, 4 (1865), 1–22; H. Brackenbury, 'Ancient Cannon in Europe. Part II: From A.D. 1351 to 1400', *Proceedings of the Royal Artillery Institution*, 5 (1867), 23–59.

6 For a recent discussion of these accounts see Thom Richardson, *The Tower Armoury in the Fourteenth Century* (Leeds: Royal Armouries, 2016), pp. 174–90.

7 Some of the surviving accounts for the late fifteenth century have been published: see H. L. Blackmore, *The Armouries of the Tower of London* (London: H. M. S. O., 1976).

8 David Grummitt, *The Calais Garrison: War and Military Service in England, 1436–1558* (Woodbridge: Boydell Press, 2008); David Grummitt, 'The Defence of Calais and the Development of Gunpowder Weaponry in England in the Late Fifteenth Century', *War in History*, 7 (2000), 253–72. For other works on Calais see John Riley Rainey Jr, 'The Defence of Calais, 1436–1477' (unpublished doctoral thesis, Rutgers University, 1987); Paul Leroy Holmer, 'Studies in the Military Organization of the Yorkist Kings' (unpublished doctoral thesis, University of Minnesota, 1977).

Maritime historians have also investigated the use of gunpowder weapons in naval warfare by the English. They have identified the late fifteenth century as a period in which the numbers of guns on ships increased dramatically but have debated the effectiveness of these weapons in combat.[9] Other areas that have been explored include the history of early artillery fortifications.[10] John Kenyon has argued that two main phases of technological change took place, in c.1380–1420 and later in 1481–1539. These periods of innovation were prompted by the threat of invasion from the Continent, with little development taking place between 1420 and 1481.

Research into the history of gunpowder weapons possessed by English towns has been carried out by Gunn, Grummitt and Cools.[11] They have demonstrated the heavy investment by the inhabitants of these settlements in the late fifteenth to sixteenth centuries in firearms. Specific studies of individual towns have also been undertaken by historians such as Stoyle, who has highlighted the extensive use of guns by the citizens of Exeter in the Tudor period.[12] Some general works on specific reigns or conflicts also include valuable discussions of English gunpowder

9 For the argument that guns on ships were of limited value in naval warfare before the sixteenth century see Ian Friel, 'Oars, Sails and Guns: The English and War at Sea, c.1200–c.1500', in *War at Sea in the Middle Ages and Renaissance*, ed. John B. Hattendorf and Richard W. Unger (Woodbridge: Boydell & Brewer, 2003), pp. 69–80; Ian Friel, *The Good Ship: Ships, Shipbuilding and Technology in England, 1200–1520* (London: British Museum Press, 1995). For a contrary view see Kelly DeVries, 'The Effectiveness of Fifteenth-Century Shipboard Artillery', *The Mariner's Mirror*, 84 (1998), 389–99; Geoffrey Parker, 'The "Military Revolution," 1560–1660 – a Myth?', *The Journal of Modern History*, 48 (1976), 195–214 (pp. 89–90).

10 See especially John R. Kenyon, 'Early Artillery Fortifications in England and Wales: A Preliminary Survey and Reappraisal', *Archaeological Journal*, 138 (1981), 205–40; For other works on fortifications see B. H. St J. O'Neil, *Castles and Cannon: A Study of Early Artillery Fortifications in England* (Oxford: Clarendon Press, 1960); Kelly DeVries, 'Facing the New Technology: Gunpowder Defenses in Military Architecture Before the Trace Italienne, 1350–1500', in *The Heirs of Archimedes: Science and the Art of War through the Age of Enlightenment*, ed. Brenda J. Buchanan (London: Massachusetts Institute of Technology, 2005), pp. 37–72; Allen Brown, *English Castles* (Woodbridge: Boydell Press, 2004), p. 129; M. W. Thompson, *The Decline of the Castle* (Cambridge: Cambridge University Press, 1987), pp. 35–7.

11 Steven Gunn, David Grummitt and Hans Cools, *War, State and Society in England and the Netherlands, 1477–1559* (Oxford: Oxford University Press, 2006).

12 Mark Stoyle, *Circled with Stone: Exeter's City Walls, 1485–1660* (Exeter: University of Exeter Press, 2003). See also Richard Howlett, 'Norwich Artillery in the Fourteenth Century', *Norwich, Norfolk and Norwich Archaeological Society*, XVI (1907), 46–75; G. J. Mayhew, 'Rye and the Defence of the Narrow Seas: A 16th-Century Town at War', *Sussex Archaeological Collections*, 122 (1984), 107–26; Randall Moffett, 'The Military Organization of Southampton in the Late Medieval Period, 1300–1500' (unpublished Ph.D. thesis, University of Southampton, 2009); Helen Clarke, *Sandwich: The*

weapons at that time.[13] Similarly, studies of European artillery or discussions of the military revolution debate frequently refer to the use of guns in England but are often restricted by a lack of chronological context and a reliance on a narrow body of printed primary sources.[14] Therefore, little effort has been made to explore the overall development of gunpowder weapons in England over an extended period.[15]

Yet an abundance of surviving documentary sources means that this topic can be explored in detail.[16] This is particularly the case with the financial records of the

"Completest Medieval Town in England": A Study of the Town and Port from Its Origins to 1600 (Oxford: Oxbow, 2010).

13 See Chris Given-Wilson, *Henry IV* (New Haven: Yale University Press, 2016), pp. 454–5; Anthony Goodman, *The Wars of the Roses: Military Activity and English Society, 1452–97* (London: Routledge, 1981), pp. 159–72. See also C. T. Allmand, *Henry V* (New Haven: Yale University Press, 1997), pp. 215–16; R. R. Davies, *The Revolt of Owain Glyn Dŵr* (Oxford: Oxford University Press, 1995), p. 253; Jonathan Sumption, *The Hundred Years War: Volume IV, Cursed Kings* (London: Faber, 2015), pp. 418–19.

14 Bert S. Hall, *Weapons and Warfare in Renaissance Europe* (London: The John Hopkins Press Ltd, 1997). See also Philippe Contamine, *War in the Middle Ages*, trans. Michael Jones (Oxford: Blackwell, 1984), pp. 142, 148; M. G. A. Vale, *War and Chivalry: Warfare and Aristocratic Culture in England, France and Burgundy at the End of the Middle Ages* (London: Duckworth, 1981), p. 146; Kelly DeVries, 'The Impact of Gunpowder Weaponry on Siege Warfare in the Hundred Years War', in *The Medieval City under Siege*, ed. Ivy A. Corfis and Michael Wolfe (Woodbridge: Boydell Press, 1995), pp. 227–44; Kelly DeVries, 'The Technology of Gunpowder Weaponry in Western Europe During the Hundred Years War', in *Xxii. Kongreß Der Internationalen Kommission Für Militärgeschichte Acta 22: Von Crécy Bis Ohács Kriegswesen Im Späten Mittelalter (1346–1526)* (Vienna: Heeresgeschichtliches Museum, 1997), pp. 285–98; Kelly DeVries, *Medieval Military Technology* (Peterborough, Ontario: Broadview Press, 1992); Maurice H. Keen, 'The Changing Scene: Guns, Gunpowder, and Permanent Armies', in *Medieval Warfare: A History*, ed. Maurice Keen (Oxford: Oxford University Press, 1999), pp. 273–92; Clifford J. Rogers, 'The Military Revolutions of the Hundred Years' War', *The Journal of Military History*, 57 (1993), 241–78 (pp. 276–7); Geoffrey Parker, *The Military Revolution: Military Innovation and the Rise of the West, 1500–1800* (Cambridge: Cambridge University Press, 1988); Anne Curry, 'Guns and Goddams: Was there a Military Revolution in Lancastrian Normandy, 1415–50?', *Journal of Medieval Military History*, 8 (2010), 171–88 (pp. 187–8); Simon Pepper, 'Aspects of Operational Art: Communications, Cannon, and Small War', in *European Warfare, 1350–1750*, ed. Frank Tallett and D. J. B. Trim (Cambridge: Cambridge University Press, 2010), pp. 181–202. For the origins of the military revolution debate see Michael Roberts, *Essays in Swedish History* (London: Weidenfeld & Nicolson, 1967).

15 For a full discussion of the historiography on this topic see Dan Spencer, 'The Development of Gunpowder Weapons in Late Medieval England' (unpublished doctoral thesis, University of Southampton, 2016), pp. 9–20.

16 For a full discussion of the sources for this topic see ibid., pp. 3–9.

English crown, of which the most substantial quantity is made up of documents from the Exchequer. This was the principal royal treasury of the English crown, with the Lower Exchequer of Receipt responsible for receiving income and issuing money for payments, whereas the Upper Exchequer audited royal revenues.[17] Additional information is available from the office of the Chancery. This was a large office headed by the Chancellor, responsible for keeping enrolments of important letters in the Royal Chancery, including the Patent and Close Rolls.[18] Other important sources held at The National Archives include the records of the Duchy of Lancaster and a series of artificially created classes of documents known as Special Collections, which include many miscellaneous estates that were periodically held by the crown.[19] A large body of financial accounts also survives for English towns although, by comparison with the records of the crown, they are far less complete and comprehensive.[20] Most of these royal and urban accounts remain unpublished in manuscript form.[21] These documents provide plentiful information on the purchase and repair of guns, the construction of artillery fortifications, the wages of gunners and numbers of guns. However, these sources rarely provide specific details on how guns were used or why they were acquired.

This limitation can be rectified, to a limited extent, by making use of narrative sources. Chroniclers, such as the anonymous author of the *Gesta Henrici Quinti*,

17 A. L. Brown, *The Governance of Late Medieval England 1272–1461* (London: Edward Arnold, 1989), pp. 52–3.

18 Ibid., p. 45.

19 For more information see *Records of the Duchy of Lancaster*, http://discovery.nationalarchives.gov.uk/details/r/C83 [Accessed 11 April 2018]; *Records of Various Departments, Arranged Artificially According to Type, and Formerly Entitled Special Collections*, http://discovery.nationalarchives.gov.uk/details/r/C246 [Accessed 11 April 2018].

20 For a full discussion of these sources see Chapter 6 and Spencer, 'The Development of Gunpowder Weapons in Late Medieval England', pp. 7–9.

21 Although some important documents relating to the privy wardrobe, the king's ships, the chancery and a small number of towns have been printed. For examples see Tout, 'Firearms in England in the Fourteenth Century', pp. 666–702; Susan Rose, ed., *The Navy of the Lancastrian Kings: Accounts and Inventories of William Soper, Keeper of the King's Ships, 1422–1427* (London: Allen & Unwin for the Navy Record Society, 1982); M. Oppenheim, ed., *Naval Accounts and Inventories of the Reign of Henry VII, 1485–8 and 1495–7, Publications of the Navy Records Society* (London: Naval Records Society, 1896); *Calendar of the Close Rolls* (London: H. M. S. O., 1900–63); *Calendar of the Patent Rolls* (H. M. S. O., 1900–); *Fourth Report of the Royal Commission on Historical Manuscripts* (H. M. S. O., 1874); *Fifth Report of the Royal Commission on Historical Manuscripts* (H. M. S. O., 1877–8); *Sixth Report of the Royal Commission on Historical Manuscripts* (H. M. S. O., 1879).

can provide interesting insights into the use of guns in sieges and on the battlefield.[22] Yet most English narrative sources rarely discuss gunpowder weapons in any detail. Another useful source of information is provided by the household accounts of magnates, such as for John Howard, duke of Norfolk.[23] However, these documents rarely survive in sufficient quantities to provide useful insights into gunpowder weapons held by the nobility. More problematic is the lack of any records produced by gun-makers, gunners or other workers involved in the production or use of guns. For example, manuals on gunnery were not produced in England prior to the second half of the sixteenth century.[24] The evidence available for the topic is therefore heavily skewed towards royal and urban financial accounts. This means that although information on expenditure on guns is plentiful, details of how and why they were used often needs to be inferred from a careful analysis of the sources. The wealth of evidence for guns held by the crown and the towns means that these areas are the focus of the book.

The first part of this book examines the use of guns by the English crown in land warfare, overseas expeditions, naval warfare, the Calais garrison and royal castles. The first chapter explores the major changes in royal artillery that occurred over the fifteenth century, such as the development of new gun types and the administration of the ordnance. The second chapter then serves as a case study which focuses on the ordnance used in two expeditions of the fifteenth century, in 1430–2 and 1497. These have been selected because an exceptional quantity of information survives detailing how the artillery was procured and used during those campaigns. The third chapter looks at how the use of guns on royal ships developed in the fifteenth century. The records for the garrison of the strategic territory known as the Pale of Calais are explored in the fourth chapter. The survival of extensive sources for this enclave means that long-term developments in guns can be traced in detail from the reign of Richard II to the beginning of the reign of Henry VII. Chapter 5 looks at English castles and the way guns were used throughout the period.

The second part of the book investigates the use of artillery by English towns. The experience of English towns in the adoption and use of guns, including the role of royal policy in driving these changes, is examined in the sixth chapter. The seventh chapter then serves as a case study by exploring the sizeable extant records for the town of Southampton. Part three focuses on the guns themselves, with the evidence collected in the previous chapters being used in Chapter 8 to assess how

22 For example, for the siege of Harfleur see Frank Taylor and John. S. Roskell, eds., *Gesta Henrici Quinti* (Oxford: Oxford University Press, 2004), pp. 27–51.

23 Anne Crawford, ed., *The Household Books of John Howard, Duke of Norfolk, 1462–1471, 1481–1483* (Stroud: Sutton for Richard III & Yorkist History Trust, 1992).

24 Steven A. Walton, 'The Art of Gunnery in Renaissance England' (unpublished doctoral thesis, University of Toronto, 1999), p. 74.

ordnance was constructed, repaired, operated and transported. The intention of this approach is to provide both a chronological and thematic structure for the subject, through an in-depth examination of the available evidence to explore how and why royal and urban guns changed from their introduction to the end of the reign of Henry VII.

Part I

Royal

1

Royal Guns on Land

MILITARY CAMPAIGNS IN THE fifteenth century were characterised by wars with the traditional enemies of England, principally France and Scotland, as well as with rebellious subjects. The long running conflict with France known as the Hundred Years' War began in 1337 with the attempted confiscation of the duchy of Aquitaine by the French king Philip VI. Three years later, in 1340, Edward III asserted his claim to the French throne which, despite being given up at the Treaty of Brétigny in 1360, was also claimed by his successors. Henry V asserted this right to the French throne in 1415 and launched his first invasion of France. As a result of his military victories, including the conquest of much of Normandy from 1417 to 1419, he was able to negotiate the Treaty of Troyes with Charles VI in 1420, which recognised him as the latter's heir, but he died in 1422 before becoming king. He was succeeded by his infant son Henry VI, with the conduct of the war against the supporters of the Dauphin, later Charles VII of France, directed by the young Henry's uncles, the dukes of Bedford and Gloucester. Henry VI was crowned king of France in 1431 but the French were later able to drive the English out of all of their lands in France, with the exception of the Pale of Calais, by 1453. The loss of almost all of England's Continental possessions did not end Anglo-French hostilities, however, with major expeditions being sent to France in the reigns of Edward IV and Henry VII. The Franco-Scottish alliance, known as the Auld Alliance, meant that conflict between England and Scotland was also linked to war with France. For much of the century, hostilities largely took the form of limited cross-border raids, but large English armies invaded Scotland in 1400, 1482 and 1497.[1]

The political instability which resulted from Henry Bolingbroke's usurpation of the throne in 1399 meant that the fifteenth century was also defined by rebellions and civil war. Henry IV was forced to contend with major uprisings in Wales, led by Owain Glyndŵr, and in the north of England, led by the powerful Percy family.[2] The ramifications of the English defeat in France and the weakness of the

1 For a general account of the Hundred Years' War see, Anne Curry, *The Hundred Years' War: 1337–1453* (Oxford: Osprey, 2002).

2 For the Owain Glyndŵr rebellion see R. R. Davies, *The Revolt of Owain Glyn Dŵr* (Oxford: Oxford University Press, 1995); for the Percy rebellions see J. L. Kirby, *Henry*

government of Henry VI led to the outbreak of the Wars of the Roses following the first battle of St Albans in 1455, with the king's cousin Richard, duke of York, claiming the throne. York was killed at the battle of Wakefield in 1460, but his son, Edward, earl of March, was crowned king in 1461 and decisively defeated his Lancastrian enemies in the same year at the battle of Towton. The first half of his reign saw frequent Lancastrian rebellions, however, and he was briefly deposed from 1470 to 1471. Richard, duke of Gloucester, was able to usurp the throne after the death of his brother in 1483 but was in turn defeated by Henry Tudor at the battle of Bosworth in 1485. Despite this victory, Henry VII was forced to contend with rebellions for much of his reign, principally focused on the pretenders Lambert Simnel and Perkin Warbeck.[3]

The English crown lacked the resources to finance a permanent standing army and the raising of forces for campaigns was largely dependent on parliamentary grants of taxation. Expeditionary armies in the fifteenth century were recruited using the indenture system, which had developed in the previous century, whereby captains indented with the crown to provide a specified number of soldiers for a certain length of time. The vast majority of these men consisted of men-at-arms and archers although small numbers of specialist artificers, such as carpenters, masons and gunners, were also recruited. Changes did occur to the composition of armies in this period, with the proportion of archers increasing relative to the number of men-at-arms, as well as the advent of billmen in the late fifteenth century.[4] However, the most significant development in English warfare in the fifteenth century was the growing importance of gunpowder artillery, as is evidenced by an increase in the number of guns and gunners, as well as from their widespread deployment in sieges, in naval warfare and on the battlefield. This chapter explores how and why the artillery used in expeditions changed over time.

EDWARD III TO RICHARD II

The earliest recorded use of guns during an English expedition occurred on 22 August 1346 at the battle of Crécy. According to the Florentine chronicler, Giovanni Villani, Edward III deployed his army around carts that were equipped with guns whose shots during the fighting inflicted heavy losses on the French and were 'so

---

IV of England (London: Constable, 1970).

3 For a general account of the Wars of the Roses see Anthony Goodman, *The Wars of the Roses: Military Activity and English Society, 1452–97* (London: Routledge, 1981).

4 Anne Curry, 'English Armies in the Fifteenth Century', in *Arms, Armies and Fortifications in the Hundred Years War*, ed. Anne Curry and Michael Hughes (Woodbridge: Boydell Press, 1994), pp. 39–68 (pp. 41–5); David Grummitt, *The Calais Garrison: War and Military Service in England, 1436–1558* (Woodbridge: Boydell Press, 2008), p. 121.

loud and threatening that it seemed as if God himself was thundering'.[5] However, this appears to have been an isolated experiment, as there is no evidence of their use on the battlefield by the English again before the middle years of the fifteenth century.[6] Instead guns were used exclusively in siege warfare as both offensive and defensive weapons, as at the siege of Calais, which commenced in September 1346. On 1 September, Edward III ordered the delivery of ten guns with lead shot, gunpowder, sulphur and saltpetre to be transported to the besieging army outside the town. These weapons were supplied to Edward III by the keeper of the privy wardrobe, Robert of Mildenhale, an official responsible for the production and storage of arms and armour at the Tower of London.[7] Yet for much of the reign of Edward III, guns were only produced in small quantities and were mainly supplied for the furnishing of royal castles, alongside traditional artillery such as bolt-throwing springalds.[8] The principal reason for this was English strategy in these years, which was focused on mounted raids into French territory known as *chevauchées*, as well as the small size of guns in this period and the high price of gunpowder.[9] This changed with the resumption of war with France, following the breakdown of the Treaty of Brétigny in 1369, which led to an increase in the production of guns, most likely as a reaction to English defeats on the Continent.

The reign of Richard II saw further increases in the numbers of royal guns kept at the Tower of London. This gradual rise reached its peak in 1388, with the privy wardrobe accounting for a total of ninety-eight guns, together with almost 4,000lbs of gunpowder and over 2,000 gunstones.[10] Once again this increase was

5 Other accounts are provided by Jean Froissart and the authors of the *Istorie Pistolesi* and the *Grandes Chroniques de France*. Michael Livingston and Kelly DeVries, eds., *The Battle of Crecy. A Casebook* (Liverpool: Liverpool University Press, 2015), pp. 114–19, 122–5, 130–1, 268–9, 449–50. Also see T. F. Tout, 'Firearms in England in the Fourteenth Century', *The English Historical Review*, 26 (1911), 666–702 (pp. 671, 688–90); Thom Richardson, *The Tower Armoury in the Fourteenth Century* (Leeds: Royal Armouries, 2016), p. 136; Michael Prestwich, 'The Battle of Crécy', in *The Battle of Crécy, 1346*, ed. Andrew Ayton and Philip Preston (Woodbridge: Boydell Press, 2005), pp. 139–57 (pp. 148, 154).

6 Dan Spencer, 'The Lancastrian Armament Programme of the 1450s and the Development of Field Guns', *The Ricardian*, 25 (2015), 61–70 (p. 67).

7 Tout, 'Firearms in England in the Fourteenth Century', p. 670; A. L. Brown, *The Governance of Late Medieval England 1272–1461* (London: Edward Arnold, 1989), p. 58.

8 For springalds see Richardson, *The Tower Armoury in the Fourteenth Century*, pp. 133–40.

9 For chevauchées see Michael Prestwich, *Armies and Warfare in the Middle Ages: The English Experience* (New Haven and London: Yale University Press, 1996), p. 6. For the small size of guns in the reign of Edward III see Tout, 'Firearms in England', p. 683. For the high price of gunpowder see Appendix D.

10 Tout, 'Firearms in England', pp. 676–7, 686, 696–700.

primarily motivated by the needs of defence, particularly during the invasion scare of 1385–6 when large quantities of guns were supplied to royal castles in England, Brittany and Calais.[11] At the same time, there was also an increase in the calibre of some of these weapons, which shows that a process of diversification was occurring. This can be seen from the ten guns purchased for the defence of Cherbourg in 1379, of which seven fired stones of twenty-four inches in diameter and three of fifteenth inches in diameter.[12] The adoption of larger ordnance appears to have been a reaction to the successful use of gunpowder artillery by the French at the siege of Saint-Sauveur-le-Vicomte in 1374–5.[13] There is no evidence, however, that guns were supplied from the privy wardrobe for expeditions prior to the 1390s.

Richard II did lead large armies to Ireland on two occasions, in 1395 and 1399, which were both equipped with artillery.[14] In 1394, Ralph Hatton, keeper of the privy wardrobe, supplied six large guns with 300 gunstones and six small guns with 140lbs of lead shot, and later, in 1399, provided eight guns together with 200 gunstones, 200lbs of gunpowder and 200 tampons.[15] It appears that the artillery was unused in both expeditions, as the privy wardrobe accounts do not record any expenditure of ordnance.[16] Nevertheless the presence of guns in these royal campaigns foreshadows the regular deployment of these weapons in the reign of Henry IV. By the end of the fourteenth century, therefore, guns had grown in importance in warfare and were being used in increasing numbers. Yet they were mainly deployed defensively and were rarely sent to equip expeditions.

HENRY IV

Henry IV took a personal interest in gunpowder artillery and employed guns in siege warfare on a far more regular basis than his predecessors. In this, he may have

11 Ibid., pp. 678, 696–700.

12 By contrast the guns constructed in the reign of Edward III were very small, see Tout, 'Firearms in England', pp. 682–4; TNA, E 403/475, m. 20.

13 This has been identified as the first time that guns made a decisive impact during a siege by historians such as Jonathan Sumption and Christopher Allmand. Jonathan Sumption, *The Hundred Years War: Volume III, Divided Houses* (London: Faber, 2009), p. 224; C. T. Allmand, *The Hundred Years War: England and France at War, c.1300–c.1450* (Cambridge: Cambridge University Press, 1988), p. 79.

14 Nigel Saul, *Richard II* (New Haven and London: Yale University Press, 1997), pp. 277–80, 288–9.

15 Tampons were wads made out of wood that were used to separate the gunpowder from the projectile, see H. L. Blackmore, *The Armouries of the Tower of London* (London: H. M. S. O., 1976), p. 246; Tout, 'Firearms in England', pp. 700–2; TNA, E 403/459, mm. 6, 11, 16; E 403/551, m. 4; E 403/561, m. 16.

16 Tout, 'Firearms in England', pp. 700–2.

been influenced by his experiences in central and southern Europe in the 1390s, prior to becoming king of England. In 1390 he travelled to the Baltic with a small English force which assisted the Teutonic Knights in their military operations against their Lithuanian enemies. This included taking part in an unsuccessful attack on the city of Vilnius, with the siege being abandoned after the attackers ran out of gunpowder. Afterwards Henry spent three months with the order in Königsberg.[17] The Teutonic Knights had regularly employed artillery, including large guns, in siege warfare since the 1380s.[18] It is possible that Henry's time in Prussia inspired his interest in guns, particularly as he later employed foreign gunners in his service when king. This can be seen from the annuities of £10 which he granted in 1406 to two gunners, Walter Cokke and Anthony Herman, at least one of whom appears to have been, judging by his name, of foreign origin.[19] As a result of his pilgrimage to Jerusalem in 1392, Henry also met the rulers of European states known for their use of artillery in this period, including Sigismund of Luxembourg, king of Hungary, and Antonio Venier, Doge of Venice, which may have been influential in his attitudes towards guns.[20] Further evidence for the king's interest in the technology is indicated by an entry in the accounts of the treasurer of Calais for 1432–4, which records that a large gun of bronze weighing 3,904lbs was marked with the arms of England and was cast with the words: 'Henry IV made this out of the metal of forty-nine guns.'[21]

After the accession of Henry IV to the throne in 1399, responsibility for the storage and production of royal artillery became more diversified than under Edward III or Richard II. Guns continued to be constructed at the Tower of

17 Anthony Tuck, 'Henry IV and Chivalry', in *Henry IV: The Establishment of the Regime, 1399–1406*, ed. Gwilym Dodd and Douglas Biggs (York: York Medieval Press, 2003), pp. 55–72 (59–60).

18 Eric Christiansen, *The Northern Crusades* (London: Penguin Books, 1997); Mark Whelan, 'Catastrophe or Consolidation? Sigismund's Response to Defeat After the Crusade of Nicopolis', in *Between Worlds: The Age of the Jagiellonians*, ed. Florain Ardelean, Christopher Nicholson and Johannes Presier-Kapeller (Frankfurt am Main: Peter Lang, 2013), pp. 215–27 (p. 223).

19 *CPR 1405–1408*, pp. 216, 229; Chris Given-Wilson, *Henry IV* (New Haven: Yale University Press, 2016), pp. 454–5.

20 Tuck, 'Henry IV and Chivalry', p. 62. For Sigismund of Luxembourg and guns see László Veszprémy, 'The Birth of Military Science in Hungary: The Period of the Anjou and Luxembourg Kings', in *A Millenium of Hungarian Military History*, ed. László Veszprémy and Béla K. Király (New York: Columbia University Press, 2002), pp. 26–53. For the Venetian use of guns see M. E. Mallett and J. R. Hale, *The Military Organization of a Renaissance State: Venice c. 1400 to 1617* (Cambridge: Cambridge University Press, 1984).

21 This entry was recorded in the accounts for 1432–4, which states that the gun was given by the assent of John, duke of Bedford, to Peter, count of Saint-Pol, for the siege of Saint-Valery-sur-Somme on 3 August 1433, TNA, E 364/69, rot. B dorse.

London; with the last surviving set of privy wardrobe accounts covering John Norbury's period of office for the years 1404 and 1405.[22] A Robert Walys was also appointed, on 15 November 1399, to the office of maker of bows, springalds, crossbows and guns within Pontefract Castle in the Duchy of Lancaster. A month later he was granted a further commission to take workmen and saltpetre for the king's artillery in the castle.[23] Pontefract was to remain an important location for the storage and production of guns throughout the course of the first half of the reign. Gerard Sprong, esquire, also played a significant role in the organisation of the king's artillery. He was appointed by writ of the privy seal on 23 December 1400 to receive guns and other equipment from the keeper of the privy wardrobe, John Norbury. He was also entrusted with the transport, maintenance and construction of at least some of the king's guns. Sprong's accounts, submitted in 1415, indicate that his role grew in importance over time, and may have begun to supersede that of the keeper of the privy wardrobe by the end of the reign.[24] They also reveal that a significant increase in the size of guns occurred in the early fifteenth century, as can be seen by the emergence of very heavy guns with their own individual names, such as the *Messenger* of 4,480lbs and *King's Daughter* of 5,600lbs.[25] The organisation of the royal artillery, therefore, underwent significant changes under Henry IV. This was also reflected in the increased use of guns in siege warfare in campaigns in Scotland, Wales and the north of England.

Henry Bolingbroke, earl of Derby, had been exiled by his cousin Richard II in 1397, but taking advantage of the latter's absence in Ireland, he invaded England and took possession of the realm in the summer of 1399. Richard was taken prisoner and forced to abdicate, with Henry being crowned in October. Less than a year later, in August 1400, the new king invaded Scotland with a large army. This was intended to deter Scottish raiding into England and to try to force the Scots to recognise his overlordship over Scotland.[26] The artillery provided for the expedition consisted of the eight guns previously sent by Richard II to Ireland, in addition to twenty-four quarrel guns, saltpetre and sulphur.[27] These guns are unlikely to have been used, however, as the English army only spent eight days in Scotland and achieved very little.[28]

22 Richardson, *The Tower Armoury in the Fourteenth Century*, pp. 11, 231; TNA, E 101/404/25.

23 *CPR 1399–1401*, pp. 100, 171.

24 TNA, E 364/49, rot. C.

25 See Appendix A.

26 Anne Curry, 'New Regime, New Army? Henry IV's Scottish Expedition of 1400', *The English Historical Review*, 125 (2010), 1382–1413 (p. 1,382).

27 TNA, E 403/564, m. 13; E 403/567, mm. 1, 11, 16.

28 Curry, 'New Regime, New Army?', p. 1393.

Artillery was also deployed on a limited scale in the early years of the Owain Glyndŵr rebellion in Wales. This had begun as a small uprising in Ruthin in the north-east in September 1400, but spread to most of Wales over the next few years, posing a major threat to English rule.[29] On 26 April 1401 a commission was granted to William Woodward, a founder of London, and to Gerard Sprong, to take bronze, copper, charcoal and saltpetre for the making and working of guns in London and elsewhere. Sprong was later instructed to transport artillery to Wales on 22 September.[30] In the following year he also delivered 100lbs of saltpetre and 50lbs of sulphur on 16 August for a royal expedition to Wales.[31] The limited provision of artillery in this period was dictated by the nature of the guerrilla warfare, with its emphasis on raids as opposed to formal sieges, although small quantities of guns and gunpowder were provided for the defence of royal castles.[32]

From 1404 onwards the provision of guns became a more important aspect of campaigns in Wales and elsewhere, as a reaction to rebel victories, notably the capture of the major castles of Harlech and Aberystwyth.[33] This can be seen from the 1,000lbs of saltpetre and 600lbs of gunpowder given to Sprong in February 1404, together with the 8,000lbs of iron given to him in April of the following year, for the repair of guns in the Tower of London.[34] Pontefract Castle was also the scene of much activity in 1404, with guns moved there and efforts made to acquire saltpetre.[35] For 1405, detailed information exists for at least part of the artillery intended for the royal expedition to Wales. On 26 April, Henry Somer, keeper of the privy wardrobe, was instructed by warrant of the privy seal to deliver ordnance to Sprong for a campaign against the rebels. This included four guns, 2000lbs of saltpetre, 500lbs gunpowder, 400lbs of sulphur with charcoal, gunstones and other equipment. The warrant also mentions the earliest recorded ordnance company in English service, made up of thirteen men, consisting of Sprong, two masons, four gunners and six carpenters, with an unspecified number of servants.[36] This was intended to support the activities of two large armies, one led by Henry, Prince of Wales, and another commanded by Henry IV.[37] The provision of artillery for this

29 R. R. Davies, *The Revolt of Owain Glyn Dŵr* (Oxford: Oxford University Press, 1995), p. 102.

30 *CPR 1399–1401*, pp. 480, 554.

31 TNA, E 364/36, rot. G.

32 TNA, SC 6/774/14, mm. 4d, 6d; SC 6/774/15, m. 1d; SC 6/775/3, m. 2d.

33 Davies, *The Revolt of Owain Glyn Dŵr*, pp. 116–17.

34 TNA, E 101/404/25; J. L. Kirby, ed., *Calendar of Signet Letters of Henry IV and Henry V (1399–1422)* (London: H. M. S. O., 1978), p. 71.

35 *CPR 1401–1405*, pp. 433, 504.

36 Kirby, *Calendar of Signet Letters of Henry IV and Henry V*, pp. 76–7; *CPR 1405–1408*, p. 60.

37 Davies, *The Revolt of Owain Glyn Dŵr*, p. 119.

campaign suggests that siege warfare was anticipated, most likely for the re-capture of key locations, such as the castles of Aberystwyth and Harlech.

In the end, Sprong appears not to have travelled to Wales as he was instead diverted to serve in the north of England against the rebellious earl of Northumberland, Henry Percy. The latter had seized an envoy of the king, Robert Waterton, in early May and had garrisoned his castles, including Warkworth and Alnwick. Northumberland's rising was seriously compromised, however, by the defeat of his co-conspirators, Richard Scrope, archbishop of York, and Thomas Mowbray, earl of Nottingham.[38] This meant that Henry IV could direct his resources against the earl's supporters in the far north of England. On 29 May 1405, Sprong was instructed to take guns from the Tower of London to the king's presence, with at least six guns subsequently moved from Pontefract Castle to Nottingham and then onwards to the north.[39] In addition to this, 5,000lbs of gunpowder was transferred from the privy wardrobe for the expedition.[40] The artillery was subsequently used at the sieges of Warkworth and Berwick castles and played an important role in their capture. According to a report sent to the council, the king had summoned the garrison of Warkworth Castle to surrender but as the defenders were well provisioned they had refused. Henry responded by attacking the castle with guns which prompted the garrison to yield on 1 July.[41] Similarly, the chronicler Thomas Walsingham stated that the garrison of Berwick Castle surrendered after a large gun had been fired at them, which demolished part of one of the towers and 'so frightened the garrison that they preferred voluntarily to expose themselves to the king's swords rather than to wait for a second firing'.[42] At least part of the ordnance was damaged in the course of the sieges, with two bronze guns broken and an unspecified number of iron guns requiring repairs in August of the same year, in addition to the expenditure of 950lbs of gunpowder.[43]

The success of Henry IV in defeating the earl of Northumberland meant that greater attention could be directed to the war in Wales. In 1406 the English achieved a number of victories against the rebels, including the capture of Anglesey. These

[38] James Hamilton Wylie, *History of England Under Henry the Fourth, Vol. II. 1405–1406* (London: Longmans, Green and Co., 1894), pp. 178, 228; Given-Wilson, *Henry IV*, pp. 266–7.

[39] *CPR 1405–1408*, p. 60; TNA, DL 42/16, f. 83v.

[40] TNA, E 101/404/25.

[41] Kirby, *Calendar of Signet Letters of Henry IV and Henry V*, p. 95; Wylie, *History of England Under Henry the Fourth, Vol. II. 1405–1406*, pp. 258–9; Given-Wilson, *Henry IV*, p. 270.

[42] David Preest trans. and James G. Clark, ed., *The Chronica Maiora of Thomas Walsingham 1376–1422* (Woodbridge: Boydell Press, 2005), p. 339; Wylie, *History of England Under Henry the Fourth, Vol. II. 1405–1406*, pp. 264–5, 270–3; Given-Wilson, *Henry IV*, pp. 270–1.

[43] TNA, E 364/49, rots. C–C dorse; *CPR 1405–1408*, p. 64.

successes meant that the reduction of the Welsh-held castles of Aberystwyth and Harlech could now be carried out in earnest in the years 1407 to 1409.[44] The preparations for these sieges included the transport of guns, gunpowder and gunstones by land from Hereford and by sea via Bristol.[45] In 1408, expenditure on the construction of guns at the Tower of London totalled £106 6s. 8d., including one which was said to have been 'newly invented by the Lord the King himself'.[46] A further £368 was spent on saltpetre, sulphur and bows in 1407 and 1408.[47] The extensive provision of artillery suggests that these sieges were intended as showcases to demonstrate English might; instead the beleaguered Welsh garrisons were able to put up a prolonged and determined resistance.[48] For the siege of Aberystwyth Castle in 1407, 7,287lbs of saltpetre, 2,026lbs of sulphur and 122 large gunstones were sent from the privy wardrobe to the besiegers, whose artillery train consisted of at least five guns, four of which broke. The artillery included an iron gun called *Neelpot*, a bronze gun called *Messenger* weighing 4,800lbs and two unnamed guns of bronze weighing 500lbs. A further iron gun called *Clyff* was left at the castle after its surrender in autumn 1408. During the siege of Harlech, 2,825lbs of gunpowder was expended by the attackers with a further 5,700lbs of gunpowder placed in the stores of the captured castle. At least three named guns were used at the siege, including the *King's Daughter*, which broke and which may have been designed by Henry IV himself, and two other guns called *God's Grace* and *Fowler*.[49] In both sieges gunpowder artillery was ineffectual against the walls of these castles, despite the presence of large prestigious guns and substantial quantities of gunpowder, with the garrisons instead being starved into surrender, but they did demonstrate the technological superiority of the English over the rebels.[50]

Henry's victories over his domestic opponents, in Wales and the north of England, meant that he could focus more attention on dealing with France in the last years of his reign. Relations with Charles VI of France had been tense following the usurpation of Henry IV in 1399 and French armies had attacked Calais and Gascony in the years 1404–7.[51] The assassination of Louis of Orléans in 1407,

44 Davies, *The Revolt of Owain Glyn Dŵr*, pp. 122–5.

45 TNA, E 403/591, mm. 5, 8; E 404/22, no. 551; *CPR 1405–1408*, p. 359.

46 TNA, E 403/594, mm. 8, 18; E 403/595, mm. 1, 16–18; *CPR 1405–1408*, pp. 420, 473.

47 TNA, E 403/591, m. 9; E 403/594, m. 1; E 403/595, m. 21.

48 Davies, *The Revolt of Owain Glyn Dŵr*, pp. 124–5, 252–3.

49 The *King's Daughter* may have been the new large gun that was recorded as having been designed by the king in an entry dated 12 March 1408, TNA, 403/594, m. 18; E 159/184, brevia directa baronibus, Michaelmas, rot. 41; E 364/49, rots. C–C dorse.

50 Davies, *The Revolt of Owain Glyn Dŵr*, p. 253.

51 Wylie, *History of England Under Henry the Fourth, Vol. II. 1405–1406*, pp. 90–4; Guilhem Pépin, 'The French Offensives of 1404–7 against Anglo-Gascon Aquitaine', *Journal of Medieval Military History*, 9 (2011), 1–40.

however, led to the outbreak of civil war between the Armagnac and Burgundian factions, which Henry sought to benefit from by negotiating with both sides. In 1412 he agreed the Treaty of Bourges with the Armagnacs, whereby in return for the despatch of an army to assist them, they would aid him in recovering the full extent of the Duchy of Gascony. In the same year, therefore, an English army under the command of Henry's second eldest son, Thomas, duke of Clarence, was sent to France, along with two guns provided by Sprong.[52] It is unlikely that these were employed in the campaign, however, as a reconciliation of the two warring sides meant that Clarence's army was unneeded and he was paid to withdraw by the Armagnacs.[53] The reign of Henry IV, therefore, saw important changes to the administration of the royal artillery, the increased use of guns in offensive siege warfare and the naming of large guns, later referred to as bombards. Deployed against wavering garrisons, as in 1406, they appear to have been relatively useful in sieges, but against more determined opponents, as in 1407–9, they were less effective; particularly as they had a tendency to break under the strain of firing. Henry IV himself clearly took a personal interest in these weapons and they appear to have been used as a means of projecting royal power and might.

HENRY V

Henry V became king in 1413 and soon afterwards initiated preparations for a major expedition to France, which included a large artillery train. He clearly appreciated the importance of guns in siege warfare, despite his own mixed experiences of these weapons at the sieges of Aberystwyth and Harlech Castles. This may have been due, at least in part, to the increased use of guns on the Continent in the early years of the fifteenth century.[54] The need for large quantities of ordnance led to significant changes to the organisation of the royal artillery during the course of his reign. Guns continued to be forged in the Tower of London under the supervision of the privy wardrobe up until 1419.[55] The privy wardrobe was eventually superseded, however, by the new ordnance office, with the first master of the ordnance, Nicholas Merbury, appointed on 22 September 1414.[56] Some idea of the artillery available to Henry V at the beginning of his reign can be seen from

52 Anne Curry, *Agincourt: A New History* (Stroud: Tempus, 2005), pp. 22, 29–32; TNA, E 364/49, rots. C–C dorse.

53 Curry, *Agincourt*, p. 31.

54 For the use of guns by John the Fearless, duke of Burgundy see Robert D. Smith and Kelly DeVries, *The Artillery of the Dukes of Burgundy 1363–1477* (Woodbridge: Boydell Press, 2005), pp. 76–83.

55 TNA, E 403/642, m. 3.

56 *CPR 1413–1416*, p. 241.

the accounts of Gerard Sprong, submitted in 1415, which reveal that Henry had at least six large and twenty small guns, located in the Tower of London, Wales and elsewhere.[57] Thereafter, no enrolled or particulars of account survive for ordnance in the reign of Henry V.[58] This means that there is limited information on exactly how many guns were used in expeditions in this period. Fortunately there are numerous entries in the issue rolls related to artillery which bear testament to the significant role of guns in Henry's campaigns.

A crucial element of Henry V's plans for his invasion of France was the provision of a powerful artillery train. Gunpowder weapons had been deployed on a limited scale in previous overseas expeditions, such as in 1346–7 and 1412, but the extensive resources allocated to the procurement of guns for the 1415 campaign was unprecedented in English history.[59] In 1414, efforts were made to increase the production of guns in the Tower of London, with the king's smith, William Merssh, appointed to hire smiths, and 18,000lbs of iron purchased by William Gunmaker.[60] In addition to this, £107 10s. 8d. was spent on the construction of one large gun at Bristol.[61] Saltpetre and sulphur were also transported to the Tower of London and as many as 10,000 gunstones were purchased.[62] A German or Dutch knight William Van Jsendorne was also paid £33 6s. 8d. for sourcing master gunners and gunpowder, most likely from the Low Countries.[63] This was an industrialised region which had been well known for the production of guns since the late fourteenth century.[64] Henry was undoubtedly keen to benefit from the expertise of these specialists to ensure that he had a large and powerful artillery train. Jsendorne's efforts appear to have borne fruit as, in addition to Gerard Sprong, 29 master gunners and 58 other gunners were retained for the expedition in 1415, many of whom appear to have been German or Dutch, judging by their names.[65] They were paid at a premium rate, with the master gunners receiving 20d. per day and the assistant gunners 8d. per day. Other professionals hired for the king's ordnance included 100 masons, 120 carpenters, 40 smiths and 60 carters.[66] The

57 TNA, E 364/49, rot. C dorse.

58 The exceptions to this pertain to royal ships and the Calais garrison which are dealt with in Chapter 3 and Chapter 4 respectively.

59 Tout, 'Firearms in England in the Fourteenth Century', pp. 671–2; TNA, E 364/49, rots. C–C dorse.

60 *CPR 1413–1416*, p. 177; TNA, E 403/617, m. 13.

61 TNA, E 403/614, m. 15.

62 TNA, E 403/617, m. 4; E 403/619, m. 3.

63 TNA, E 403/619, m. 2.

64 Sergio Boffa, *Warfare in Medieval Brabant, 1356–1406* (Woodbridge: Boydell Press, 2004), pp. 158–60, 190.

65 TNA, E 101/45/5, mm. 8–9.

66 *CPR 1413–1416*, p. 346.

large quantities of ordnance that were procured for the expedition demonstrate that a campaign of conquest was intended from the outset.

The expedition set sail for Normandy on 11 August and the siege of Harfleur began seven days later.[67] Artillery played an important role in the siege of the town, as can be seen in the detailed account given in the *Gesta Henrici Quinti*. According to the author of the *Gesta*, Henry positioned his guns and mechanical artillery close to the walls and towers of Harfleur. The English artillery was then able to bombard the town in relative safety, as they were protected by large wooden screens and trenches which they had dug. The French, however, were also well equipped with firearms and in front of their gates they had constructed a new specialised form of fortification known as a bulwark. These were small forts constructed out of wood and earth with embrasures for the use of guns and other missile weapons. This appears to have been the first time that the English encountered this new type of artillery fortification, which they soon adopted for their own use. The author of the *Gesta* reports that the English artillery 'by the violence and fury of the stones' caused considerable damage to the defences, in addition to other buildings, 'almost as far of the middle of the town'. This account can be corroborated by the quantities of materials which are recorded as being expended by the besiegers in a writ enrolled in the King's Remembrancer's Memoranda Roll for 1417–18. A total of 7,466 gunstones were fired by the attackers, 876 by the bombards, 3,600 by the large fowlers and 3,000 by the small fowlers. Harfleur was besieged for six weeks, with the defenders only surrendering on terms after the main bulwark had been captured.[68] The English bombardment clearly caused significant damage, as can be seen by the repairs carried out to the defences of the town after its capture on 22 September, with masons, carpenters and labourers, including French prisoners, employed on mending walls, towers and bulwarks. This was not the sole reason for the successful outcome of the siege, with the exhaustion of the garrison, the failure of the French to relieve the town and the threat of a storm also being important factors.[69] It is clear, however, that the use of artillery at Harfleur was considered by Henry to be a success, judging by the substantial investment in ordnance for his future campaigns in France. The army's guns were then left behind in the town

67 Curry, *Agincourt*, pp. 81–3.

68 Note that this is the earliest known reference to fowlers in English accounts. Frank Taylor and John. S. Roskell, eds., *Gesta Henrici Quinti* (Oxford: Oxford University Press, 2004), pp. 30–49; TNA, E 159/194, brevia directa baronibus, Michaelmas, rot. 6. By contrast, the earliest reference to a *veuglaire*, the name used by the French for a fowler, in Burgundian records occurs in 1417, Smith and DeVries, *The Artillery of the Dukes of Burgundy 1363–1477* (Woodbridge: Boydell Press, 2005), p. 231.

69 Curry, *Agincourt*, p. 97.

along with many of the gunners; Gerard Sprong commanded eighteen master gunners and eighteen assistant gunners in the garrison until at least 1420.[70]

In 1416 the English were mainly on the defensive, although over £700 was spent on the procurement of ordnance. Guns were forged in the Tower of London under the direction of William Merssh and purchased from William Woodward, a founder of London. Additional artillery was also transported to reinforce the garrison at Harfleur.[71] The most substantial expenditure on guns, however, occurred from 1417 to 1419, during Henry's conquest of Normandy. Significant sums of money were spent on gunpowder, saltpetre and sulphur: over £300 in 1417, almost £600 in 1418 and over £300 in 1419.[72] This was a campaign characterised by siege warfare, with artillery therefore playing an important role.[73] On 1 August 1417, Henry invaded Normandy with a large army and laid siege to the city of Caen seventeen days later. Caen was heavily bombarded by the English artillery before being taken by storm on 4 September. The brutal treatment of the inhabitants persuaded a number of other settlements in Lower Normandy to surrender quickly, including Argentan and Alençon, although Falaise was only captured after a two-month siege.[74] The heavy expenditure of gunpowder during these sieges meant that it was necessary for further munitions to be sent from England to France. In 1418 large quantities of supplies were transported from Southampton to Caen, including ninety-three barrels of saltpetre and seventeen barrels of gunpowder.[75] Regular payments also occurred throughout the year for the construction of guns at Tower Hill by the smith John Matthew, together with other ordnance which was shipped from Winchelsea to Caen.[76] The effective deployment of guns also required large quantities of other materials and animals, as is revealed by an entry on the patent roll for 10 February directed to John Louthe, clerk of the works of the king's ordnance. Louthe was directed to take workers for the construction of equipment for guns including 300 large pavises, 7,000 tampons, 7,000 gunstones and 12 large carts, in addition to halters and other items for 100 oxen and 320 horses to draw the same vehicles.[77]

Further towns and castles in Lower Normandy were captured in 1418, including Avranches and Cherbourg. Henry then invested the capital of Normandy, Rouen,

70 TNA, E 36/79, ff. 40–4, 56–7.

71 TNA, E 403/624, mm. 3–4, 6, 8, 11, 14.

72 TNA, E 403/630, mm. 2, 4, 13; E 403/633, mm. 1, 4; E 403/636, mm. 1, 4, 7, 9, 14; E 403/639, mm. 1–2, 6; E 403/644, m. 3.

73 Christopher Allmand, *Henry V* (New Haven: Yale University Press, 1997), p. 215.

74 Richard Newhall, *The English Conquest of Normandy 1416–1424: A Study in Fifteenth Century Warfare* (New Haven: Yale University Press, 1924), pp. 56–79.

75 TNA, E 403/636, m. 14.

76 TNA, E 403/633, mm. 1, 13; E 403/636, mm. 1, 4, 7, 9; E 403/639, mm. 2, 5; E 159/200, brevia directa baronibus, Trinity, rot. 9.

77 *CPR 1416–1422*, pp. 133–4.

on 29 July, and after a siege of twenty-four weeks the city was starved, rather than bombarded, into surrender. In the following year, military operations were aimed at consolidating English control of Normandy.[78] Additional artillery was sent over to France from England, with fowlers and other guns forged at the Tower of London.[79] The final years of the reign of Henry V continued to see some ordnance transported from England to support the war effort in France. In February 1420, John Louthe was again instructed to hire artificers for the construction of equipment including four large carts which were to be drawn by sixty oxen and twenty-four horses.[80] In October of the following year, 1,600lbs of gunpowder were purchased from William Woodward and 8,100lbs of saltpetre acquired from a merchant of Catalonia, Conrade Albert, with a further 11,036lbs of saltpetre supplied by a London merchant, Thomas Chalton, all of which were sent to Harfleur and other fortifications in February 1422.[81] Nevertheless, by the time of Henry V's death in 1422, it is clear that the production of artillery for the war was now mostly carried out in Lancastrian France, using French rather than English financial resources, as is evidenced by a sharp fall in expenditure by the English Exchequer. This seems to have occurred due to the capture of large quantities of artillery during the conquest of Normandy and as a result of the king's efforts to construct guns in Lancastrian France itself.[82] Henry's recognition as heir to Charles VI, as a result of the Treaty of Troyes in May 1420, also meant that the former acquired additional French resources, as well as an alliance with Philip, duke of Burgundy, who was himself well provided with gunpowder artillery.[83]

Henry V made extensive use of artillery in his war with France and spent large sums of money on ordnance. His employment of guns was probably influenced by his father's use of these weapons, yet he used them on a much greater scale. All subsequent major expeditions to France by his successors also deployed significant quantities of artillery. His reign can therefore be seen as representing a turning point in the use of guns by English monarchs. This was made possible, at least in part, by a significant decrease in the price of gunpowder sold in England which occurred in the early fifteenth century.[84] In the late fourteenth century the price had ranged from 12d. to 24d. per lb, whereas by 1417 the price had decreased to 5d. per lb.[85] This made

78 Newhall, *The English Conquest of Normandy 1416–1424*, pp. 94, 110–21, 129.

79 TNA, E 403/639, m. 14; E 403/642, m. 3.

80 *CPR 1416–1422*, p. 262.

81 TNA, E 403/646, m. 1; E 403/649, m. 5; E 159/198, brevia directa baronibus, Hilary, rot. 25.

82 Dan Spencer, 'The Provision of Artillery for the 1428 Expedition to France', *Journal of Medieval Military History*, 13 (2015), 179–92 (pp. 187–9).

83 Allmand, *Henry V*, pp. 140–5.

84 See Appendix D.

85 Bert Hall has also identified a fall in the price of European gunpowder in the late fourteenth century, Bert S. Hall, *Weapons and Warfare in Renaissance Europe* (London:

it much more affordable to acquire and use large quantities of gunpowder in siege warfare. The employment of guns was not a guarantee of success in capturing enemy settlements – as seen at the siege of Harfleur other factors often came into play – but guns could be useful in sufficiently damaging fortifications to make attempts at storming possible or to persuade the defenders to surrender.

HENRY VI

Henry VI was only nine months old when he became king in 1422, which meant that a council was appointed to govern for him until he came of age, led by his uncles, with John, duke of Bedford, as regent of France, and Humphrey, duke of Gloucester, as lord protector of England.[86] The war was continued against the Armagnac supporters of the Dauphin, the future Charles VII, with English armies sent on a regular basis to the Continent.[87] In contrast to much of the reign of Henry V, however, artillery was infrequently transported from England to France. Instead, guns were only dispatched to accompany major expeditions, such as in 1428, 1430, 1441 and 1443, as well as during periods of crisis, notably in 1436 and 1449–50. This was because artillery was constructed in Lancastrian France, such as at Rouen, and then distributed to different masters of the ordnance.[88] The office of master of the ordnance in England did not exist on a permanent basis; instead men were appointed on an ad hoc basis to serve for the duration of specific campaigns.[89] John Hampton, for instance, was appointed as master of the king's ordnance in 1430, with responsibility for the artillery for the royal expedition to France.[90] After the return of the king to England in 1432, however, Hampton relinquished this office and a similar official was not appointed to the position until 1449.[91] This meant that the production and administration of artillery was mostly concentrated

The John Hopkins Press Ltd, 1997), p. 64.

86 Ralph A. Griffiths, *The Reign of King Henry VI: The Exercise of Royal Authority, 1422–1461* (London: Benn, 1981), p. 17.

87 Curry, 'English Armies in the Fifteenth Century', pp. 44–5.

88 For instance see TNA, E 364/69, rot. G.

89 By contrast, masters of the ordnance appear to have been appointed on a permanent basis in France for specific geographical regions, such as for Normandy. For a discussion of the masters of the ordnance see C. T. Allmand, 'L'artillerie de l'armée Anglaise et Son Organisation a l'époque de Jeanne d'Arc', in *Jeanne d'Arc, une Époque, un Rayonnement*, ed. R. Pernaud (Paris: CNRS, 1982), pp. 73–83; Andy King, 'Gunners, Aides and Archers: The Personnel of the English Ordnance Companies in Normandy in the Fifteenth Century', *Journal of Medieval Military History*, 9 (2011), 65–75.

90 TNA, E 364/69, rot. Q; E 404/47, no. 117.

91 Hampton was described as previously being master of the ordnance by May 1432, TNA, E 403/703, m. 2.

in France, as is evidenced by the comparatively infrequent references to ordnance in the issue rolls of the English Exchequer between 1422 and 1450. This situation was transformed in 1449 with the appointment of Thomas Vaughan as master of the ordnance and the production of large numbers of guns in England.[92]

The conquest of Normandy was continued by Bedford in the first two years of the reign of Henry VI, with armies being sent from England in support of military operations.[93] It is probably for this reason that John Acclan, a yeoman of the crown, was paid 26s. 8d. in 1423 for the carriage of gunpowder and other equipment to Winchelsea. After this date, however, there is no evidence in Exchequer records for the use of guns in an English expedition before 1428.[94] Artillery continued to be important in siege warfare in France, as can be seen by the accounts of Sir Lancelot de Lisle, who served as marshal for the army of Thomas Montagu, earl of Salisbury. In 1425, Lisle received from John Harbotel, master of the ordnance for the duke of Bedford, a total of 12,800lbs of gunpowder and 4,380 gunstones for the sieges of La Guerche, Le Mans, Sainte Suzanne and Mayenne.[95] Two years later, instructions were given by the regent of France, John, duke of Bedford, for four bombards to be transported from Harfleur to Paris for the recovery of the town of Montargis.[96]

In 1428 the decision was taken to send an army from England under the command of the earl of Salisbury to attack the Loire Valley. This included a large artillery train sent with the expedition, which was assembled in England and financed by the English Exchequer. The decision to send ordnance to France was probably motivated by the need to replace guns lost at the failed siege of Montargis in the summer of 1427 and to augment the resources of Lancastrian France.[97] John Parker, master of the ordnance for the earl of Salisbury, was granted 1,000 marks on 24 March 1428. Over the next five months this was used to purchase 71 guns, consisting of 7 bombards, 48 fowlers and 16 handguns, in addition to other equipment, such as 1,214 gunstones, 18 carts, cables, pulleys and tools. The artillery was also accompanied by an ordnance company consisting of 96 men including master gunners, assistants, carpenters, masons, labourers and others. In August the ordnance was shipped to France in five ships and was then transported to the siege of Orléans. The artillery played a prominent role in the bombardment of the city which began on 17 October and by 24 October had resulted in the capture of the tower of the Tourelles. The

92 *CPR 1446–1452*, p. 332.

93 Griffiths, *The Reign of King Henry VI*, p. 185.

94 TNA, E 403/660, m. 10.

95 BL, Add Ch 17269; Mark William Warner, 'The Montagu Earls of Salisbury, 1300–1428' (unpublished doctoral thesis, University College London, 1991), pp. 172–3, 180–1.

96 I am grateful to Professor Emmanuel de Crouy-Chanel for providing this reference. BNF, Français 26050, no. 771.

97 Anne Curry, 'Guns and Goddams: Was there a Military Revolution in Lancastrian Normandy, 1415–50?', *Journal of Medieval Military History*, 8 (2010), 171–88 (p. 183).

army's ordnance was subsequently deployed in bulwarks, which were constructed around the perimeter of the town, and the guns were fired at the city throughout the course of the siege. They were also used defensively to protect these fortifications from French attacks. Much of this artillery is likely to have been captured by the French following the lifting of the siege of Orléans in May 1429.[98]

At the end of the same year, the decision was taken by the king's council to send a large expedition to France accompanied by the king himself.[99] This was a reaction to the crowning of Charles VII at Reims on 17 July and to the reverses suffered by the English in France.[100] On 24 February 1430, John Hampton, master of the ordnance, was appointed to take workers for the construction of artillery. Hampton spent over £2,000 on ordnance for the expedition, including the construction of 3 bombards, called *Henry, Hampton* and *Crown*, 20 fowlers, 1,550 gunstones, 1,200lbs of gunpowder, over 5,000lbs of saltpetre and 5,200lbs of sulphur, together with other equipment. The artillery train was accompanied by an ordnance company of 95 men, consisting of gunners, masons, carpenters, smiths, carters, and other workers. Hampton also received large quantities of ordnance from Lancastrian France, including three bombards, called *Flouredelys*, *Newgate*, and *Towerwharf of Cumbria*, nineteen fowlers, fourteen cart guns almost 8,000lbs of gunpowder and 10,000lbs of saltpetre. This is the earliest known reference in Exchequer records to English cart guns, which were later used regularly in military operations in the 1430s and 1440s.[101] Hampton subsequently distributed this equipment to different armies operating in northern France in the period 1430–2. A substantial proportion of the ordnance was expended during the course of military operations, including twenty fowlers, over 6,000lbs of gunpowder, over 5,000lbs of saltpetre, 2,700lbs of sulphur and other equipment. Thirty-one of the remaining guns were allocated to the master of the ordnance in Normandy, with the other guns and equipment distributed to castles in the duchy.[102]

Henry VI was crowned king of France in Paris at the end of 1431 but this did not lead to an improvement in English fortunes.[103] The death of Bedford and the defection of the duke of Burgundy in 1435 meant that the English were increasingly on the defensive. In the following year, Calais was besieged by a large Burgundian army with

98 Spencer, 'The Provision of Artillery for the 1428 Expedition to France', pp. 179–92; TNA, E 101/51/27.

99 For a more detailed discussion of this expedition see Chapter 2.

100 Anne Curry, 'The "Coronation Expedition" and Henry VI's Court in France, 1430 to 1432', in *The Lancastrian Court*, ed. Jenny Stratford (Donington: Shaun Tyas, 2003), pp. 29–52 (pp. 29–30).

101 Note that these are described as *ribaudequins* in these accounts. For a full discussion of the terminology of these guns see Appendix J.

102 TNA, E 364/69, rots. Q–Q dorse, R.

103 Griffiths, *The Reign of King Henry VI*, p. 193.

an impressive artillery train, while Crotoy and Cherbourg were also increasingly threatened by the French.[104] To support these beleaguered garrisons, substantial quantities of ordnance were sent out from England in the years 1435 to 1438.[105] However, the defection of the duke of Burgundy may well have caused problems for the English in their attempts to keep up with technological developments. The expertise of Burgundian gunners had previously been useful in English service, such as in 1430, but the availability of these professionals became problematic after the siege of Calais in 1436.[106] This appears to have encouraged the government of Henry VI to look further afield for specialist technological expertise. In 1438 payments were made for the expenses of a delegation of fourteen men of 'Ducheland', including doctors of theology, esquires and gunners, with the latter being paid to test fire different types of ammunition in the presence of the king in Oxford.[107] These included a range of large and small gunstones, as well as more unusual varieties, such as four in the shape of crosses, two of which were 'rabetted' with iron, as well as one iron gunstone.[108]

In the early 1440s relatively little artillery was sent out from England to support Lancastrian France. In part, this was because significant quantities of guns still remained in Normandy, as can be seen from the accounts of William Forsted, master of the ordnance for the duchy.[109] Richard, duke of York, when appointed as lieutenant-general of Normandy in 1440 did request, however, that he be provided with six large guns, twelve large fowlers and twenty-six gunners along with other supplies of ordnance.[110] In the following year, William Hickling, controller of

104 Grummitt, *The Calais Garrison*, pp. 20–31; Griffiths, *The Reign of King Henry VI*, pp. 201, 460–2.

105 TNA, E 403/721, mm. 9, 17; E 403/723, mm. 3, 11; E 403/725, mm. 7–8, 15; E 403/729, m. 8; E 403/735, m. 5.

106 For the recruitment of Burgundian gunners in 1430 see M. K. Jones, 'The Beaufort Family and the War in France, 1421–1450' (unpublished doctoral thesis, University of Bristol, 1982), p. 82.

107 The term used to describe these artificers is men of 'Ducheland', which could mean that they were from the Low Countries or German lands further to the east. As the Burgundians were at war with the English at this time it is likely that they were from the latter, TNA, E 101/503/7; E 364/74, rot. J.

108 This appears to mean that a recess was cut out of a gunstone with iron then fixed in the gap, which indicates that this was a composite stone and iron type of projectile. For a definition of rabbet see C. T. Onions, ed., *The Shorter Oxford English Dictionary*, volume 2 (Oxford: Clarendon Press, 1973), p. 1735.

109 Jones, 'The Beaufort Family and the War in France', pp. 135, 140, 142; Joseph Stevenson, ed., *Letters and Papers Illustrative of the Wars of the English in France during the Reign of Henry the Sixth, King of England*, volume 2, part 2 (London: Longman, Green, Longman, and Roberts, 1864), pp. 463–4.

110 Stevenson, *Letters and Papers Illustrative of the Wars of the English in France*, volume 2, part 2, pp. 587–8.

the ordnance in Normandy, was supplied from England with over 18,000lbs of saltpetre, 2,000lbs of sulphur and other equipment.[111] In 1443, John Dawson, master of the ordnance for John Beaufort, duke of Somerset, was allocated £1,100 for the purchase of artillery for an expedition to France. This included the construction of twenty cart guns and fifty culverins, the hiring of six gunners together with the delivery of 4,000lbs of saltpetre and 3,000lbs of sulphur.[112] The decision to procure light guns that were easy to transport in the field was dictated by the strategy of the campaign, which was to carry out a *chevauchée* in northern France that, it was hoped, would result in a decisive battle with the army of Charles VII. It may also suggest that the English were experimenting with guns that could have been potentially used on the battlefield. Yet the expedition was delayed in reaching France and achieved very little before it returned to England.[113]

Negotiations between the two sides resulted in a truce being agreed between England and France in 1444 at the Treaty of Tours. The war was resumed, however, after an English attack on the Breton town of Fougères in March 1449.[114] This led to a French invasion of Normandy which rapidly overran many of the towns and castles of the duchy, with Rouen being captured by November.[115] Henry VI's government responded by despatching sizeable quantities of ordnance to Caen and Cherbourg in late 1449 and early 1450, in an attempt to secure their remaining possessions in Lower Normandy.[116] These measures were unsuccessful, with the defeat of an English army at the battle of Formigny in April leading to the loss of the entire duchy by October. The reasons for this sudden defeat included the inadequate leadership and poor morale of the English, as well as the strength of the French armies.[117] French superiority in artillery also played an important role in their victory, as can be seen from the account of the chronicler, Gilles de Bouvier, who claimed that the king of France had such a great number of 'large bombards, large cannons, veuglaires, serpentines, crapaudeaux, ribaudequins and coleuvrines that no man can remember any Christian king ever

111 TNA, E 403/741, m. 4; E 404/57, no. 264.

112 Note that this is the earliest mention of culverins in English accounts, for a discussion of this gun type see Appendix J. *CPR 1441–1446*, p. 199; TNA, E 403/747, m. 15; E 403/749, mm. 6, 7, 8, 15; E 404/59, nos. 161, 162, 247; Harris Nicolas, ed., *Proceedings and Ordinances of the Privy Council of England*, volume 5 (London: Record Commission, 1835), p. 281.

113 Jones, 'The Beaufort Family and the War in France', pp. 162–5, 170, 173–7; Michael K. Jones, 'John Beaufort, Duke of Somerset, and the French Expedition of 1443', in *Patronage, the Crown and the Provinces*, ed. R. A. Griffiths (Gloucester: Sutton, 1981), pp. 79–102.

114 Griffiths, *The Reign of King Henry VI*, pp. 490–510.

115 Ibid., pp. 515–22; Jones, 'The Beaufort Family and the War in France', pp. 256–77.

116 TNA, E 403/777, mm. 5, 8; E 403/785, m. 4; E 403/785, m. 3; E 404/66, nos. 26, 35, 80–1, 86, 97, 177, 181; Griffiths, *The Reign of King Henry VI*, p. 518.

117 Griffiths, *The Reign of King Henry VI*, pp. 521–2.

having such great artillery.'[118] Older types of guns, such as bombards, continued to be important in the sieges of towns and castles in Normandy, but it appears that the use of a new type of gun called a serpentine was also significant. This was a long narrow gun which was first recorded as being used by the French in 1442.[119] By contrast, the earliest known evidence for the possession of serpentines by the English occurs in 1448.[120] From this point onwards, until the 1480s, they were produced and used in large numbers, on the battlefield and in siege warfare. The rapid adoption of this type of gun by the English in 1450 suggests it was a reaction to the successful deployment of serpentines by the French.

The loss of Normandy contributed to the outbreak of a major rebellion in south-east England led by Jack Cade, in the summer of 1450.[121] The king's military preparations to counter this included taking ordnance provided by Thomas Vaughan, master of the ordnance, 'in to the feeld' on 20 June, with the artillery train consisting of two serpentines, one culverin and seven large cart guns, which were accompanied by an ordnance company of twenty-five gunners, carpenters, smiths and masons.[122] This is the earliest evidence for the intended use of guns on the battlefield by an English army for the fifteenth century, although it is unlikely that they were actually deployed against the rebels as they were not needed in the end. Later in 1453 it appears that efforts were made to furnish an abortive expedition to France. On 15 May, William Hickling was appointed to take workers for the construction of guns for the use of the king in Gascony, possibly to reinforce John Talbot, earl of Shrewsbury, who was killed at the battle of Castillon in July. Hickling's accounts reveal that he spent £752 12s. 5d. on the construction of thirty-five guns, consisting of three bombards, called *Goodgrace*, *Henry* and *Crown*, four large fowlers, four serpentines and twenty-four cart guns between 1453 and 1456.[123] They also demonstrate that some guns had become considerably heavier by the 1450s, with the *Goodgrace* weighing as much as 22,400lbs, whereas the fowlers each averaged 1,960lbs.[124] Any plans for a new expedition to recover the lost lands in

118 For definitions of these gun types see the Glossary and Appendix J, note that the English equivalent of a *veuglaire* was a fowler; Joseph Stevenson, ed., *Narratives of the Expulsion of the English from Normandy, M. CCCC.XLIX.–M.CCC.L* (London: Longman, Green, Longman, Roberts, and Green, 1863), p. 373.

119 Emmanuel de Crouy-Chanel, *Canons Médiévaux: Puissance du Feu* (Paris: Rempart, 2010), p. 78.

120 MCO, FP 43, f. 13r.

121 For a discussion of Cade's rebellion see Griffiths, *The Reign of King Henry VI*, pp. 610–16.

122 TNA, E 404/66, no. 181; Spencer, 'The Lancastrian Armament Programme of the 1450s and the Development of Field Guns', p. 73.

123 Ibid., pp. 63–4.

124 These weights were considerably heavier than for the equivalent types of guns procured for the 1428 expedition to France see Appendix A.

France, however, were thwarted by the mental illness of Henry VI and the civil war that followed the First Battle of St Albans in 1455.[125] In 1456, the Lancastrian court decided to withdraw from London to the West Midlands and to build up their stockpile of artillery in anticipation of renewed conflict with the supporters of Richard, duke of York. On 21 December 1456, John Judde, a merchant of London, was appointed master of the king's ordnance for having constructed sixty serpentines and gunpowder to the weight of twenty tons. Twenty-six of these guns were sent to Kenilworth from London in January 1457, with Judde being paid £133 8s. 5½d. for them together with gunpowder, saltpetre and sulphur in March of the same year. It seems that the ordnance constructed by Hickling had also been transported to Kenilworth, as he was ordered to make payment for the carriage of twenty-one cart guns and four serpentines 'nowe being at oure castell of Kenelworth and taken by Judde unto oure house of oure ordinaunce at oure citee of London ther to be stokked and carted'.[126]

Therefore, by the end of the 1450s the government of Henry VI had ample stocks of guns. Part of this stockpile had been dispersed for coastal defence in 1459, but the Lancastrians are still likely to have had a sizeable artillery train at the battle of Northampton in 1460. This is borne out by the testimony of Abbot John Whethamstede of St Albans, who states that the king had constructed a fortified camp to protect his army outside the town, with the Lancastrians rejecting negotiations prior to the engagement due to their confidence in their engines of war and the strength of their fortifications.[127] Yet during the battle itself the Lancastrians were at a disadvantage due to treachery within their ranks and to the poor weather conditions, which prevented them from using their guns. According to the author of *An English Chronicle* 'The ordenaunce of the kynges gonnes avayled nat, for that day was so grete rayne, that the gonnes lay depe in the water, and so were queynt and myghte nat be shott.'[128] The Yorkists, perhaps in response to the efforts made by the Lancastrians, also had their own field guns, which suggests that these weapons were considered essential for use on the battlefield by this time. They supposedly had guns during the first battle of St Albans in 1455, and at the battle of Ludford Bridge in 1459 they were said to have fortified their position and

125 Spencer, 'The Lancastrian Armament Programme of the 1450s and the Development of Field Guns', p. 64.

126 Ibid., pp. 64–5; *CPR 1452–1461*, p. 342; TNA, E 404/71/1, no. 60; E 404/71/3, no. 1.

127 H. T. Riley, ed., *Registra Quorundam Abbatum Monasterii S. Albani, Qui Sæculo XVmo. Floruere. Vol. I, Registrum Abbatiae Johannis Whethamstede, Abbatis Monasterii Sancti Albani, Iterum Susceptae; Robert Blakeney, Capellano, Quondam Adscriptum* (London: Longman & Co., 1872), pp. 373–4.

128 John Sylvester Davies, ed., *An English Chronicle of the Reigns of Richard II, Henry IV, Henry V, and Henry VI, Written Before the Year 1471* (London: Camden Society, 1856), pp. 96–7.

set carts with guns in front of their army. Following their victory at the battle of Northampton, the Yorkists made efforts to increase their stock of guns, with £143 13s. 4d. being allocated for this purpose in December. In the same month, the duke of York and earl of Salisbury were accompanied by an artillery train in their march north, which culminated in the battle of Wakefield.[129] By the end of the reign of Henry VI, therefore, significant changes had occurred to the types and ways that guns were deployed in warfare. The king himself seems to have shown very little personal interest in guns or in promoting their usage, but they continued to be important in the prosecution of the war in France and later during the battles of the Wars of the Roses. Significant sums of money were spent on supplying large quantities of ordnance for the major campaigns, although changes in the types and ways that guns were used in warfare were relatively limited before the 1450s. The most important development was the adoption of serpentines, which were then rapidly used in sizeable numbers on the battlefield and in siege warfare.

EDWARD IV

The reign of Edward IV saw significant changes in the numbers, types and organisation of the royal guns, particularly after 1471. The office of the master of the ordnance was established on a permanent basis for the first time, with the master assisted by a yeoman and clerk of the ordnance.[130] Extensive efforts were also made to acquire new ordnance for the garrison of the Pale of Calais, with the victualler of Calais, William Rosse, playing an important role in the purchasing of artillery for the 1475 expedition to France. This was due in part to Edward's keen interest in guns, as can be seen by his involvement in the preparations for the French expedition, with an Italian observer stating that 'every day he inspects all his artillery…and although he has a large number of bombards, he has fresh ones made every day.'[131] The Anglo-Burgundian alliance also resulted in the transmission of new gun types and specialist personnel to England. The expedition accounts for the 1475 and 1481 campaigns, together with the plentiful entries relating to ordnance in the issues and tellers' rolls, strongly suggests that English artillery underwent important changes in the reign of Edward IV.

129 Spencer, 'The Lancastrian Armament Programme of the 1450s and the Development of Field Guns,' p. 66.

130 By contrast it appears that previous masters of the ordnance in the 1450s had been unpaid, as can be seen by a petition made by Thomas Vaughan in 1455, TNA, SC 8/145, no. 7217.

131 A. R. Myers., ed., *English Historical Documents, 1327–1485*, volume 4 (London: Eyre & Spottiswoode, 1969), p. 526.

Edward, earl of March, became the Yorkist claimant to the throne after the death of his father, Richard, duke of York, at the battle of Wakefield in December 1460. Edward won a decisive victory over the Lancastrians at the battle of Towton in March 1461 and was crowned three months later.[132] The new king continued to face challenges to his rule, however, and was forced to contend with rebellions in Wales and the north of England in the early 1460s. Scottish raids in support of the Lancastrians prompted the despatch of ordnance to the border in 1461, with gunners being paid for the carriage of gunpowder to the north on 2 July.[133] At the same time, Sir William Herbert was tasked by the king with capturing Lancastrian held castles in Wales, with guns sent from Chester for the siege of Chirk Castle in August.[134] Herbert was able to defeat a Lancastrian army outside Caernarfon and to capture a number of castles, including Pembroke and Denbigh.[135]

Despite these victories in Wales, substantial threats to Edward's rule remained, with the landing of Margaret of Anjou, wife of Henry VI, with a small French army in October 1462, leading to the outbreak of a rebellion in Northumberland. Edward IV made preparations for a large army to travel to the north to confront the Lancastrians, but his illness meant that military operations were directed by his cousin, Richard Neville, earl of Warwick.[136] In November, gunners were paid to ride to Durham and other places in the north with gunpowder in the service of the king.[137] John Paston, in a letter written in December, reported that ordnance from Newcastle, mostly likely shipped there from the Tower of London, had been distributed 'bothe for the segys and for the feld'. These artillery pieces were used for the sieges of the castles of Alnwick, Dunstanburgh and Bamburgh, with field guns also being made available in case the Scots had decided to confront the Yorkists in battle.[138] In February of the following year, six gunners received wages for having served in the north, with a further payment made for one bombard of Warkworth having been transported to the siege of Bamburgh Castle.[139] Rebellion again broke

132 For a discussion of the gunshot found at Towton see Glenn Foard and Anne Curry, *Bosworth 1485: A Battlefield Rediscovered* (Oxford: Oxbow Books, 2013), pp. 146–9; Charles Ross, *Edward IV* (London: Eyre Methuen, 1974), pp. 130–41.

133 Ibid., p. 46; TNA, E 403/823, mm. 1, 3.

134 TNA, SC 6/798/7, m. 10d; SC 6/779/10, m. 6; Jane Laughton, *Life in a Late Medieval City* (Oxford: Windgather Press, 2008), p. 36.

135 Ross, *Edward IV*, pp. 48–9.

136 Ibid., pp. 50–1.

137 TNA, E 403/827A, mm. 5, 8; Cora L. Scofield, *The Life and Reign of Edward the Fourth: King of England and of France and Lord of Ireland*, volume 1 (London: Longmans, Green, 1923), p. 262.

138 James Gairdner, ed., *The Paston Letters: A.D. 1422–1509*, volume 4 (London: Chatto & Windus, 1904), p. 60.

139 TNA, E 403/827A, m. 12; Scofield, *The Life and Reign of Edward the Fourth*, volume 1, p. 264.

out in the north in 1463, with a large bombard being transported from Chester to Newcastle in March of the following year, and in April efforts were made to acquire carts for carriage of the king's ordnance.[140] On 25 June, Bamburgh Castle was once more besieged by an army commanded by Warwick, with a substantial artillery train being deployed against the walls which included the guns *Newcastle*, *London* and *Dijon*. These were able to inflict serious damage on the defences of the castle, which prompted the garrison to surrender.[141]

Artillery was later transported from the county of Chester for the expedition of John Tiptoft, earl of Worcester, to Ireland in 1465.[142] The next few years also saw increased efforts to capture the Lancastrian-held castle of Harlech in Wales. On 22 April 1466, John Wode, master of the ordnance, was paid £20 for the purchase and carriage of guns for the siege of the castle.[143] In 1467, artillery was moved from the north to the Tower of London; these guns were then transported to Harlech by sea the following year, along with ordnance from Bristol. It is likely that large numbers of guns and much gunpowder were used in the siege, as John Wode was paid over £1,300 for his expenses on ordnance in 1468 alone.[144] This included the employment of an ordnance company of sixty-seven gunners, smiths, carpenters and other professionals.[145] In a writ of 28 February 1468 these activities were described as concerning 'gret matiers of charge that we have comaunded him which touchyng the conveyaunce guyding and settyng of oure ordinaunce for a greet enterpryse whyche by godds mercy shall hastily take effect to the wele of this oure royaume'.[146] The royal army was eventually able to force the surrender of the garrison of Harlech on 14 August.[147]

These victories in England and Wales meant that Edward felt secure enough to intervene in Continental affairs, particularly in the rivalry between Louis XI, king of France, and Charles the Bold, duke of Burgundy. This led to a treaty between England and Burgundy, which included the marriage of Charles to Edward's sister, Margaret of York.[148] Anglo-Burgundian plans for a joint invasion of France were disrupted by an uprising against Edward led by his brother, George, duke of Clarence, and the earl

140 TNA, SC 6/780/2, m. 1d.

141 Ross, *Edward IV*, p. 61; Myers, *English Historical Documents, 1327–1485*, volume 4, pp. 289–90.

142 *CPR 1461–1467*, p. 488.

143 TNA, E 403/835, m. 1; E 101/329/2.

144 TNA, E 403/838, m. 6; E 403/840, mm. 4, 11, 18; E 404/74/1, nos. 68, 90; PSO 1/29, no. 1518.

145 TNA, E 404/74/1, no. 67.

146 TNA, E 159/245, brevia directa baronibus, Pascha, rot. 18d.

147 Scofield, *The Life and Reign of Edward the Fourth*, volume 1, p. 459.

148 Richard Vaughan, *Charles the Bold: The Last Valois Duke of Burgundy* (rpt Woodbridge: Boydell Press, 2002), pp. 47–8.

of Warwick. Edward was forced to flee the country and take exile in Burgundy in late 1470, whilst Henry VI was briefly restored to the throne.[149] The former invaded England in the following year, however, landing in the north with a small force, which was said to have included 300 Flemings armed with handguns.[150] After marching southwards, Edward's army confronted Warwick's at the battle of Barnet on 14 April 1471, with both sides employing field artillery. According to the newsletter of Gerhard von Wesel, Warwick's serpentines and handguns were fired continuously the night before the engagement, but their shot fell short of their opponents. On the day of the battle, Edward's army won a decisive victory, with his guns said to have won the artillery duel with their opponents.[151] At the same time, however, Prince Edward, son of Henry VI, and his mother, Margaret of Anjou, landed in Weymouth, and joined a Lancastrian army that had been raised in the south-west. Edward therefore made preparations to confront his opponents, which, according to the author of the *Arrivall of Edward IV*, included 'all things that was thowght behovefull for a new field … so purveyed he artilary, and ordinaunce, gonns, and othar, for the filde gret plentye'. The two armies met at the battle of Tewkesbury on 4 May, where the Yorkists' superiority in artillery appears to have played a role in their victory. The king's ordnance was said to have been 'conveniently layde afore them', which 'sore oppressyd' their Lancastrian enemies, who had fewer guns, and provoked the latter to attack.[152] Following the battle of Tewkesbury, Edward headed eastwards to London, whose inhabitants had been besieged by an army led by the Bastard of Fauconberg. However, the latter had lifted the siege of the capital prior to the return of the king to the city on 21 May.[153] Instead, Edward proceeded into Kent to pacify the county, accompanied by a strong force, which included an ordnance company of forty gunners.[154]

The deaths of Edward's principal opponents, such as the earl of Warwick, Prince Edward of Westminster and Henry VI, meant that the Lancastrian threat was effectively eliminated by the end of 1471. Yet this had been a costly victory for the royal artillery, as can be seen from a letter given under the privy seal by Edward IV to John Wode, dated 20 November 1471, which states that 'oure ordeunce

149 Ross, *Edward IV*, p. 145–55.

150 James Orchard Halliwell, ed., *A Chronicle of the First Thirteenth Years of the Reign of King Edward the Fourth by John Warkworth, D. D. Master of the St Peter's College, Cambridge* (London: Camden Society, 1839), p. 13.

151 Hannes Kleineke, 'Gerhard von Wesel's Newsletter from England, 17 April 1471', *The Ricardian*, 16 (2006), 63–88 (pp. 77–8).

152 John Bruce, ed., *Historie of the Arrivall of Edward IV. In England and the Finall Recouerye of His Kingdomes From Henry VI A.D. M.CCCC.LXXI* (London: Printed for the Camden Society by J. B. Nichols and Son, 1838), pp. 24, 29.

153 C. F. Richmond, 'Fauconberg's Kentish Rising of May 1471', *The English Historical Review*, 85 (1970), 673–92 (pp. 681–2).

154 TNA, E 403/844, mm. 4–5.

and artelerie now of late daies by oure great Enuemyes have be taken awane and spoyled and the residue therof broke and brosed in such feldes as we of late had'. The letter goes onto say that 'thoo(s) whiche remaigne untaken and unspoyled be not hable sufficient nor redy to helpe us at oure nede for the defense of our persone nor of this oure royine'. Wode was thereby commanded to purchase all 'necessaries in any wise nedefull or containing to our said ordeunce and artelerie as by your discrescon now at this tyme to be repayred or now made shalbe thought expedient for the said defense of our personne and of this our royine'. These instructions led to Wode carrying out a complete overhaul of the king's artillery in England, with work starting twelve days later, on 2 December. Wode's surviving accounts record that fifty-four serpentines and six large fowlers were refurbished by nine smiths and six wheelwrights, with a large bastard gun called the *Long Edward* being forged by another team of smiths.[155]

In 1473 John de Vere, earl of Oxford, with limited French support was able to capture St Michael's Mount in Cornwall but was effectively isolated on the island. He was later forced to surrender after being besieged by an army and fleet, with John Wode paid over £100 for the transport of artillery from the Tower of London for the siege.[156] This meant that Edward could focus on planning an invasion of France, in conjunction with the dukes of Brittany and Burgundy, to punish Louis XI for his support for the Lancastrians and to assert his claim to the French throne. Edward's preparations included assembling an impressive artillery train for the campaign, with roughly £6,000 spent on the purchase of guns, gunpowder, the wages of gunners and other expenses on ordnance between 1474 and 1476.[157] William Rosse, the victualler of Calais, was granted over £2,700, which was used to purchase forty-five serpentines as well as hundreds of gunstones and thousands of lead shot.[158] In addition to this, John Sturgeon, master of the king's ordnance, was given over £2,000 for his expenses on artillery in 1474. The same year he delivered to William Rosse thirteen large guns, including bombards, bombardelles, curtows, pot guns, fowlers and bastard guns, in addition to other equipment.[159] The procurement of new gun types for the expedition demonstrates that it was intended to be equipped

155 TNA, E 101/294/21; E 101/294/19; Dan Spencer, 'The Tower of London and Firearms in the Reign of Edward IV', *Arms and Armour*, 13:2 (2016), 98–110.

156 Ross, *Edward IV*, p. 192; Cora L. Scofield, *The Life and Reign of Edward the Fourth: King of England and of France and Lord of Ireland*, volume 2 (London: Longmans, Green, 1923), p. 85; TNA, E 405/57, m. 1; *CPR 1467–1477*, p. 412.

157 TNA, E 405/59, mm. 7r, 8r–8d; E 405/60, m. 4r; E 101/55/4, ff. 30r, 31r–31v.

158 David Grummitt, 'The Defence of Calais and the Development of Gunpowder Weaponry in England in the Late Fifteenth Century', *War in History*, 7 (2000), 253–72 (p. 265); TNA, E 101/55/4, ff.30r, 31r–31v.

159 For a definition of these guns see the Glossary and Appendix J. TNA, E 101/198/13, ff. 32r–32v; E 101/55/7.

with the most up-to-date equipment that was available. An ordnance company of over 600 men was also assembled, including thirty-six gunners, thirty-two smiths, sixty carpenters and other workers.[160] In total there were at least thirteen large guns and forty-five serpentines available for the campaign, in addition to the large quantities of artillery which were stored in Calais, accompanied by a large number of specialist personnel. Despite the extensive preparations and ambitious aims of the expedition, the invasion itself proved to be an inglorious affair. The army had crossed to Calais by early July, but the lack of support from Burgundy and Brittany meant that Edward soon came to terms with Louis in mid-August, after only very limited military operations.[161] Therefore it appears that these guns were unused, but the campaign demonstrated Edward's desire to possess weaponry to rival that of other European rulers. One Italian observer, Salvator de Clarices, went so far as to claim that 'The artillery which the King of England is bringing is greater than that of Mons. of Burgundy, which is almost incredible.'[162]

Edward later had the opportunity to showcase his large artillery train during the war with Scotland in the early 1480s. A treaty had been concluded between England and Scotland in October 1474, but relations worsened in 1479, as a result of Scottish raids, which led to the outbreak of war. This prompted Edward to undertake preparations to invade Scotland with a large royal army in 1481.[163] As with the 1475 expedition, this was to be accompanied by a sizeable quantity of guns, with roughly £3,800 being spent on ordnance from 1480 to 1482.[164] On 13 May 1480, Edward ordered the carriage of guns from Nottingham Castle to Norham Castle, which were accompanied by gunners who were paid £40 for their wages.[165] The same year 40 guns and 6 handguns were purchased from merchants by John Sturgeon for £18 4s., with a further £100 being spent on the purchase of bronze guns, bows and repairs to other guns.[166] In 1481, additional payments were made for the purchase of guns, including some from the Low Countries, hackbuts and repairs to damaged guns.[167] This also included money paid for the wages of handgunners who were serving in the north.[168] William Rosse was allocated £404 7s. 4d. for the purchase of 30 serpentines

160 Paul Leroy Homer, 'Studies in the Military Organization of the Yorkist Kings' (unpublished doctoral thesis, University of Minnesota, 1977), pp. 71–4.

161 Ross, *Edward IV*, pp. 223, 226–30; Vaughan, *Charles the Bold*, pp. 348–9.

162 Ibid., p. 220.

163 Ibid., pp. 278–80.

164 TNA, E 405/68, m. 4d; E 405/69, mm. 1r–1d, 3d; E 405/566, f. 91r; E 405/70, mm. 1r–1d, 6d, 7d.

165 *CPR 1476–1485*, p. 213; TNA, E 405/68, m. 4d.

166 TNA, E 405/68, m. 4d; E 405/69, m. 3d.

167 For a definition of a hackbut see Appendix J. TNA, E 405/566, ff. 29, 35, 39, 52, 55, 60–3, 82, 91.

168 TNA, PSO 1/50, no. 2551B.

from smiths and 68 guns from Adriano gunner of Antwerp, together with other ordnance.[169] The equipment later transported from Calais to the north of England included 7 large guns, 26 serpentines, 24 hand culverins and 111 hackbuts, sent in June 1481.[170] The ordnance company for the campaign included 400 artificers receiving wages in Newcastle, in addition to 4 gunners for the king's great guns, 3 German gunners assembled at Calais and an unspecified number at Sandwich.[171]

In the end, Edward IV's illness meant that the invasion was postponed until the following year, although a fleet under the command of John, Lord Howard, raided the east coast of Scotland.[172] An English army did cross the border into Scotland in the summer of 1482, but it was led by the king's brother, Richard, duke of Gloucester.[173] The large artillery train was accompanied by 4 master gunners and 16 other gunners, who were employed for 56 days in 1482.[174] In addition a large number of horses were needed for transporting the ordnance, with the duke of Gloucester paid £200 for this on 30 June 1482, with a further £100 allocated for acquiring 120 draught horses.[175] The principal outcome of this campaign was the successful capture of Berwick, which was bombarded by the English artillery.[176] At least some of the ordnance used in this siege was damaged, with the gun, the *Great Edward of Calais*, later requiring repairs.[177] Therefore over the course of Edward IV's reign English artillery changed significantly, especially after 1471. Large sums of money were spent on acquiring artillery, with perhaps as much as £10,000 being spent on ordnance for the French and Scottish expeditions. Edward had a personal interest in guns and was keen to obtain specialist expertise and technology from abroad. These efforts meant that his successors inherited a substantial quantity of artillery, the organisation of which was well established under the direction of the master of the ordnance and his officials. In 1483, Richard duke of Gloucester usurped the throne from his nephew, Edward V, and was crowned Richard III. His reign was short-lived, however, as he was defeated and killed at the battle of Bosworth two years later. The victor at Bosworth, Henry Tudor, had previously spent thirteen years in exile in Brittany, but returned to England in 1485 with French support.[178] An artillery train accompanied Richard III on campaign and

169 TNA, E 364/119/36, rot. C.

170 TNA, E 101/198/13, ff. 6r, 8r.

171 TNA, PSO 1/49, no. 2526A; E 405/566, ff. 29, 38, 68–9.

172 Scofield, *The Life and Reign of Edward the Fourth*, volume 2, pp. 314–15, 321–2.

173 Ross, *Edward IV*, p. 289.

174 TNA, E 405/60, m. 1.

175 TNA, PSO 1/52, no. 2697.

176 Scofield, *The Life and Reign of Edward the Fourth*, volume 2, pp. 321–2.

177 TNA, E 101/55/4, f. 23.

178 S. B. Chrimes, *Henry VII* (New Haven and London: Yale University Press, 1999), pp. 17–49.

the royal guns were deployed at the battle, but they appear to have been largely ineffectual in determining its outcome.[179]

HENRY VII

Henry VII took a personal interest in guns, influenced in part as a result of his time spent in exile in Brittany and France, which led to new types being adopted by the English, as well as the widespread use of iron gunshot. The office of master of the ordnance became more prestigious, as can be seen by the appointment of the wealthy landowner Sir Richard Guildford to the position in September 1485.[180] According to a patent given the following year, Guildford was to be assisted by twelve gunners in receipt of royal wages, in addition to a yeoman and a clerk.[181] Technological changes also occurred, with new methods of iron production developed for the creation of iron gunshot, with the earliest evidence of this occurring in East Sussex in 1490.[182] In addition to this, the primary type of metal used for guns changed from wrought iron, as had previously been the case for most of the fifteenth century, to bronze guns during his reign.[183] Bronze had traditionally been much more expensive than wrought iron, which suggests that Henry was influenced by developments in French artillery in this period.[184] The limited data available for this period also indicates that a significant decrease in the price of bronze occurred in the 1490s.[185] Furthermore, his reign saw the beginnings of a domestic saltpetre production industry due to royal encouragement; prior to this England had been solely dependent on imports.[186] Efforts were made to recruit foreign specialists from the Continent, including the Low Countries, France and Italy. From 1494 onwards, between seven and nine gunners received annual wages

179 For a discussion of the use of guns at Bosworth, see Foard and Curry, *Bosworth 1485*.

180 Sean Cunningham, 'Guildford, Sir Richard (c.1450–1506)', *Oxford Dictionary of National Biography, Oxford University Press*, 2004; online edn, Jan. 2008 http://www.oxforddnb.com/view/article/11723 [accessed 1 April 2015].

181 Goodman, *The Wars of the Roses*, pp. 160–1; *CPR 1485–1494*, pp. 77–8.

182 Brian Awty and Christopher Whittick, 'The Lordship of Canterbury, Iron-Founding at Buxted, and the Continental Antecedents of Cannon-Founding in the Weald', *Sussex Archaeological Collections*, 140 (2002), 71–81 (p. 71).

183 For an illustration of this compare the accounts for the 1475 and 1492 expeditions, TNA, E 101/55/7; E 364/119/36; E 405/78, mm. 28d, 31d, 36r.

184 Evidence for this can be seen with the employment of Bretons in the Tower of London in the late 1480s and early 1490s in the production of bronze ordnance, TNA, E 36/124, f. 39r.

185 See Appendix B.

186 David Cressy, *Saltpeter: The Mother of Gunpowder* (Oxford: Oxford University Press, 2013), p. 42; TNA, E 404/79, no. 38; E 404/80, no. 245.

from the crown for the remainder of the century, resulting in a limited degree of continuous service.[187] Artillery was used extensively in the wars of his reign, particularly in campaigns in France from 1489 to 1492 and in Scotland in 1497. The campaigns of 1497 are especially well documented due to the survival of the accounts of Robert Clifford, master of the ordnance.[188] These demonstrate that significant resources were devoted to the production and maintenance of artillery at the end of the fifteenth century.

Despite his victory at Bosworth, the new king was threatened by pretenders backed by foreign powers, although he was able to defeat the army of Lambert Simnel at the battle of Stoke in 1487.[189] It is unclear if guns were used at Stoke, but they played a prominent role in the four expeditions that were sent to Brittany and France between 1489 and 1492.[190] This was as a result of Henry VII's involvement in the Breton Wars, in an attempt to stop the semi-independent duchy of Brittany falling under French control. The treaty of Redon was agreed in early 1489, whereby Henry agreed to send military assistance to support Anne, duchess of Brittany, with alliances also being concluded with Maximilian, King of the Romans, and Isabella and Ferdinand, rulers of Castile and Aragon.[191] Therefore, a large army was assembled, accompanied by a substantial artillery train, with over £3,000 spent on ordnance between 1488 and 1492.[192] In 1488, Richard Guildford, master of the ordnance, was paid £120 in part payment of over £1,800 allocated for artillery for the expedition. A further £100 was given to Guildford for guns to protect the north of England against Scottish aggression.[193] In March 1489, part of the English army was landed in Brittany and took part in military operations against French forces.[194] The preparations for the war against France were, however, temporarily disrupted by the outbreak of a rebellion in Yorkshire in April 1489.[195] The measures taken to suppress the rising included the despatch of twelve falcons, six of which were said to have come from overseas, two serpentines and a last of gunpowder to the north on 30 April.[196] This shows that new types of guns were being imported into

187 TNA, E 405/79, mm. 8d, 10d, 22r, 38r; E 405/80, mm. 10d, 20r.

188 TNA, E 36/8.

189 For the battle of Stoke see Goodman, *The Wars of the Roses*, pp. 100–8.

190 For the Brittany and French campaigns see J. M. Currin, '"The King's Army into the Partes of Bretaigne": Henry VII and the Breton Wars, 1489–1491', *War in History*, 7 (2000), 379–412.

191 Currin, 'The King's Army into the Partes of Bretaigne', pp. 381–2.

192 TNA, E 405/75, mm. 38r, 48r; E 405/77, m.6r; E 405/78, mm. 1r–1d, 2d, 6d, 19r–20r, 23r, 25r, 28d, 31d, 34d–36r, 38r, 39r–39d, 43r, 47d, 50r.

193 TNA, E 405/75, mm. 38r, 51d.

194 Currin, 'The King's Army into the Partes of Bretaigne', pp. 390–2.

195 M. J. Bennett, 'Henry VII and the Northern Rising of 1489', *The English Historical Review*, 105 (1990), 34–59 (p. 37).

196 TNA, E 404/80, no. 68.

England, in particular light field guns called falcons, which were primarily used on the battlefield.[197]

The suppression of the rebellion in May meant that attention could again be focused on the war against France, with as many as 100 gunners being recruited to serve under John Troughton and ordnance being purchased for the defence of the castle and town of Le Conquet in Brittany.[198] At least £100 was allocated for the making of curtows, some of which were constructed by Bretons working in the Tower of London.[199] This was a type of large gun first mentioned in English records in 1475, but which was used in far greater numbers in the reign of Henry VII.[200] Further curtows, gunpowder and ammunition were sent to Brittany in 1490, along with a second expeditionary force.[201] Artillery was also purchased for the defence of the north of England, with £73 6s. 8d. expended on equipping Berwick and Carlisle. This continued into 1491, with additional ordnance being purchased for Berwick, Scarborough and Newcastle.[202] The reverses suffered by the Anglo-Breton forces in the same year, however, led to Anne's decision to come to terms with Charles VIII and the removal of the remaining English soldiers in the duchy. In 1492 Henry VII invaded France with an army and laid siege to Boulogne, but was unable to capture the town.[203] His master of the ordnance, Richard Guildford, was allocated £300 for procuring artillery for this enterprise, in addition to guns left over from the Brittany campaign.[204] Part of Guildford's accounts survive for the expedition and reveal that seventy-nine guns, consisting of two bombardelles, three curtows, twenty demi-curtows, eight serpentines, forty-six falcons and eighty hackbuts were supplied for the army, in addition to 79,200lbs of gunpowder, as well as thousands of iron, stone and lead shot. Some of the ordnance was damaged or expended during the siege, including one curtow, four demi-curtows, one serpentine and sixty hackbuts, but interestingly, none of the falcons. This was in addition to 415 iron shot and 780 gunstones, with the remaining equipment returned to the Tower of London.[205] It is noticeable that the equipment in the inventory matches the types of ordnance employed in Brittany and France at this time.[206] This can also be seen from

197 See the Glossary and Appendix J for a discussion of this gun type.

198 Currin, 'The King's Army into the Partes of Bretaigne', p. 386; TNA, E 405/78, m. 1r.

199 TNA, E 405/78, mm. 1d–2, 3, 19r; E 36/124, ff. 36v, 39r, 48r.

200 See the Glossary and Appendix J for a discussion of this gun type.

201 TNA, E 405/78, mm. 19r–20r, 23r, 25r; Currin, 'The King's Army into the Partes of Bretaigne', pp. 401–3.

202 TNA, E 405/78, mm. 28d, 31d, 36r.

203 Currin, 'The King's Army into the Partes of Bretaigne', pp. 407–12.

204 TNA, E 405/68, m. 43r.

205 Note that these accounts are damaged so only some of the information has survived, TNA, E 36/15, ff. 8r, 9r, 25r, 33v.

206 For an inventory of all the guns in Brittany in 1492 see Louis Arthur, *Le Complot Breton de M.CCCC.XCII. Documents Inédits Publiés avec Notes et Introduction par A. de La*

Francesco Guicciardini's account of the victorious French army that invaded Italy in 1494, which used bronze guns, firing iron gunstones, on carriages drawn by horses.[207]

A complete surviving inventory of the artillery in the Tower of London survives from 1495, compiled by Robert Clifford, master of the ordnance, upon his entering the office in that year.[208] This shows that a variety of ordnance were kept in the arsenal there, consisting of ninety-four bombardelles, curtows, demi-curtows, serpentines, falcons, organ guns and pot guns.[209] In addition to this, thirty-one guns were described as 'lying about the Tower' and are likely to have been deployed defensively in the walls, towers and bulwark.[210] Other equipment included seven and a half lasts of gunpowder (18,000lbs), over 2,000 shot of iron, almost 2,000 gunstones and over 1,700 lead shot.[211] In addition, there were many carts, horse collars, broken guns, cranes, moulds, tools and other items.[212] At least part of this equipment was sent to Ireland to combat the supporters of the pretender Perkin Warbeck. The ordnance was used to support the army in the lordship, which from 24 June 1494 to 24 June 1495, comprised of a force of 653 men including ninety gunners.[213]

The defeat of his partisans in Ireland prompted Warbeck to flee to Scotland in November 1495, where he was well received by James IV and recognised as Richard IV of England.[214] This led to Henry's decision to send a large expedition to Scotland in 1497, with Robert Clifford, master of the ordnance, being allocated £1,000 for the procurement of artillery in addition to handguns purchased from Flanders.[215] A total of forty-eight guns were transported to Berwick and Newcastle in twenty-three ships, comprising bombardelles, curtows, demi-curtows, serpentines and falcons, 181 hackbuts and large quantities of other equipment.[216] Additional artillery was obtained from Berwick Castle and Calais, with repairs being carried out to guns constructed in the reign of Edward IV, including the *Great Edward*, *Little Edward*

Borderie (Nantes: La Société Des Bibliophiles Bretons, 1884), p. XXXIII. For French artillery in this period see Philippe Contamine, 'L'Artillerie Royale Francaise á la Veille des Guerres d'Italie', *Annales de Bretagne*, 71 (1964), pp. 221–61.

207 Hall, *Weapons and Warfare in Renaissance Europe*, p. 159.

208 TNA, E 36/8, ff. 90r–91r.

209 Ibid., ff. 90r, 93r, 95r.

210 Ibid., f. 90v.

211 Ibid., f. 91r.

212 Ibid., ff. 91r–95r.

213 Ian Arthurson, *The Perkin Warbeck Conspiracy, 1491–1499* (Stroud: Sutton, 1997), pp. 113–16; BL, Royal MS 18 C XIV, ff. 124–6.

214 Arthurson, *The Perkin Warbeck Conspiracy*, pp. 121–2.

215 Note that this campaign is discussed in more detail in Chapter 2; TNA, E 405/79, m. 15d.

216 TNA, E 36/8, ff. 1r–6v; M. Oppenheim, ed., *Naval Accounts and Inventories of the Reign of Henry VII, 1485–8 and 1495–7* (London: Naval Records Society, 1896), pp. 104–7.

and *Messenger*.[217] Only 27 guns and 50 hackbuts were transported into Scotland with the English army on 16 August, however, with most of the large guns being left behind at Berwick. These weapons were used to capture the Tower of Ayton as well as other peel towers and were later deployed during the confrontation with the Scottish army at Halidon Hill.[218] In the same year artillery was also deployed against the Cornish rebels at the battle of Blackheath. An ordnance company of 49 gunners and 187 other professionals was retained by the king to serve for a month against the rebels of Cornwall and Devon, with 8 falcons later being transported by land from the Tower of London to Blackheath field.[219] The reign of Henry VII, therefore, saw further changes in artillery, notably the introduction of new gun types, technological developments and the production of saltpetre in England for the first time.

CONCLUSION

The fifteenth century was a period of significant change for the artillery used in English armies, but this was not a consistent process, with an increase in the rate of development in the second half of the century. In the reigns of Edward III and Richard II, guns were mainly deployed defensively and were rarely used for expeditions. This changed under Henry IV with artillery playing a more prominent offensive role in siege warfare. Henry V continued this trend with significant resources being devoted to the production of ordnance and the employment of foreign professionals. Technological changes to guns themselves, however, were relatively limited in this period, although a dramatic reduction in the price of gunpowder meant that it could be used in much greater quantities than had hitherto been possible. Artillery began a process of diversification in the reign of Henry VI, with limited changes occurring in the 1440s, such as the use of cart guns, but the most notable development was the adoption of the serpentine in the 1450s. The last quarter of the fifteenth century saw the greatest rate of change as can be seen by comparing the inventories of 1475 and 1492, with marked differences in gun types, metallurgy and ammunition. This meant that the artillery available to Henry VII, and to a lesser extent Edward IV, differed substantially from that which had been available to their predecessors, with guns being used in new ways, on the battlefield, during sieges and in naval warfare. Guns therefore had to be created in different sizes and weights to perform different tactical roles. The new terminology for gun types also appears to reflect efforts to standardise the production of artillery. This is indicated by the 1497 accounts, which state the

217 TNA, E 36/8, f. 26r.

218 Ibid., f. 9r; Arthurson, *The Perkin Warbeck Conspiracy*, pp. 174–5; Oppenheim, *Naval Accounts and Inventories of the Reign of Henry VII*, pp. 129–31.

219 TNA, E 36/8, ff. 76v–83v.

weight of ammunition and gunpowder by shot for the different guns and which correspond to different classes. For instance all of the demi-curtows required 25lbs of gunpowder to fire iron balls weighing 40lbs.[220] The enhanced importance of artillery in warfare meant that increasingly large quantities of gunpowder were required for campaigns, particularly in the 1480s and 1490s (see Table 1). In part this was due to the substantial number of guns allocated to expeditions in the reigns of Edward IV and Henry VII.[221] It was also because the new types of guns had a much greater rate of fire than had previously been possible: in 1425, 3,000lbs of gunpowder and 800 gunstones were expended at the siege of Le Mans over a two-and-a-half-week period; by contrast, in 1497, 2,250lbs of gunpowder and 1,150 lead shot were employed at the siege of four peel towers in one day alone.[222] The substantial quantities of gunpowder needed for warfare at the time, as shown by the 79,200lbs procured for the Boulogne campaign of 1492, are likely to have placed major strains on the traditional method of importing saltpetre from abroad.[223] Henry VII therefore felt obliged to issue commissions in 1489 and 1492 for royal officials to take houses, land, materials and workers for the production of saltpetre.[224]

Royal influence played an important part in allowing these changes to happen. Most fifteenth-century English kings had a clear interest in guns and were keen to promote their usage. In this, some were influenced by their experiences in countries known for their expertise in gunpowder artillery, such as with Henry IV, Edward IV and Henry VII. Direct royal involvement with ordnance is often difficult to trace, but the issue roll for 1408 states that Henry IV was responsible for designing a large gun.[225] Guns were also on occasion fired in the presence of the sovereign, sometimes for demonstrations, as in 1438, or to mark the beginning of a new campaign such as in 1497.[226] The interest of kings was not just due to practical military reasons, however, but also because gunpowder artillery became an expression and symbol of royal power, particularly when used against rebels, such as at the battle of Blackheath in 1497.[227] The prestige of the larger guns is attested

220 TNA, E 36/8, ff. 1r–6v.

221 At least 58 guns were procured for the 1475 expedition to France, in addition to the large quantities of ordnance stored at Calais. In 1492, 79 guns and hackbuts were shipped to France and a further 58 guns and 100 hackbuts to Scotland in 1497; TNA, E 101/55/4, ff. 30r–31v; E 101/55/7; E 36/15, f. 8r; E 36/8, ff. 1–9r.

222 BL, Add Ch 17529; Arthurson, *The Perkin Warbeck Conspiracy*, p. 174.

223 This does not appear to have led to an increase in the price of gunpowder, however, see Appendix D.

224 These commissions provide the earliest evidence of saltpetre production in England; *CPR 1485–1494*, pp. 320, 395.

225 TNA, E 403/594, m. 18.

226 TNA, E 101/503/7; E 364/74, rot. J; E 36/8, f. 78r.

227 Note that there had been a long tradition of naming large siege engines in the Middle Ages. In 1304, for instance, Edward I had deployed stone throwing engines at the

TABLE 1 Gunpowder and saltpetre allocated for sieges or expeditions.

Date	Gunpowder (all types)	Saltpetre	Note
1405	500	2,000	Expedition to Wales
1408–9	8,525		Siege of Harlech
1425	12,800		Sieges in France
1430–2	9,846	13,199	Expedition to France
1481	12,000		Shipped to Scotland from Calais
1484	9,800		Siege of Hammes Castle
1492	79,200		Shipped to France from the Tower
1497	15,300	32,000	Shipped to Scotland from the Tower

Source: J. L. Kirby, ed., *Calendar of Signet Letters of Henry IV and Henry V (1399–1422)* (London: H. M. S. O., 1978), pp. 76–7; BL, Add Ch 17269; TNA, E 364/49, rot. C; E 364/69, rots. Q–Q dorse; E 101/198/13, ff. 6r–6v, 20r; E 36/15, f. 8r; E 36/8, ff. 1r–7r.

by the individual names given to them, with guns often being named after the sovereign or other leading figures of the time. This was recognised by contemporary chronicle writers, who sometimes referred to these weapons by name, such as with the *Newcastle*, *London* and *Dijon* used in the siege of Bamburgh Castle in 1464.[228] The development of the office of the ordnance was also a direct result of royal interest in improving how the king's artillery was organised and administered. The most important part played by English monarchs in the progression of the technology in England, however, was in the acquisition of specialist expertise from mainland Europe. For much of the period gunners were recruited from the industrialised Low Countries, although by the reign of Henry VII they also included Bretons and Italians. The knowledge and skills of these men appear to have been crucial in the transmission of technological ideas and techniques, such as new types of guns, methods of refining gunpowder and ammunition. The defection of the duke of Burgundy in 1435 may even have contributed towards England lagging technologically behind the French in the 1440s, as it became harder to acquire Burgundian gunners. In some cases, however, it appears that technological

siege of Stirling called *Segrave*, *Vicar*, *Parson*, *Warwolf* and *Gloucester*, J. Bradbury, *The Medieval Siege* (Woodbridge: Boydell Press, 1992), p. 268.

228 Myers, *English Historical Documents, 1327–1485*, volume 4, pp. 289–90.

advances occurred without royal encouragement, such as the emergence of large guns at the beginning of the fifteenth century and the adoption of the serpentine in the 1450s. Nevertheless, royal interest in guns combined with the expertise of foreign gunners meant that English artillery kept pace with technological advances in mainland Europe for most of the period.

2

The Expeditions of 1430–2 and 1497

Carefully organised and planned expeditions often involved substantial expenditure by English kings and were essential to the conduct of warfare. This chapter offers in-depth examinations of three expeditions for which an exceptional level of detail survives. These consist of the 1430 royal-led expedition to France and two campaigns in 1497, one against Scotland and another against the rebels of Cornwall.[1] They not only provide insights into how ordnance was procured and used in these specific military operations, but also suggest significant developments that occurred in the period between the expeditions.

It is possible to carry out a comprehensive study of the ordnance used in these years due to the chance survival of complete accounts for the expeditions. These documents are available in the National Archives as the campaigns were financed by the English Exchequer. The accounts for the military operations of 1430–2 exist in the foreign account rolls of the Exchequer on three sides of membranes, enrolled in the year 1434–5.[2] They were submitted by John Hampton, master of the ordnance for Henry VI, and cover the period 24 February 1430 to 9 February 1432. The particulars of account which were used to compile Hampton's enrolled account no longer survive, but the account itself contains plentiful information on the artillery used in the expedition. This follows a standard format: the opening explanation of dates covered and the accounting official (Hampton), his receipt of money, the equipment received, the purchase of artillery, actual use of ordnance and where it was distributed at the end of the period. It therefore contains more information on how the artillery was used on the campaign than a similar set of accounts for the 1428 expedition to France.[3]

1 Henry VII did not take part in the expedition to Scotland due to the outbreak of the rebellion in Cornwall.

2 It was not uncommon for accounts to be submitted years after they had finished, for instance the accounts of the treasurer of Calais for 1413–17 were eventually included in the roll for 1434–5, TNA, E 364/69, rots. L–L dorse; E 364/69, rots. Q–R.

3 See TNA, E 101/51/27; Dan Spencer, 'The Provision of Artillery for the 1428 Expedition to France', *Journal of Medieval Military History*, 13 (2015), 179–92.

The accounts for the expeditions of 1497, by contrast, are far more voluminous as the particulars survive in a book of 110 folios.[4] The book was compiled under the direction of Robert Clifford, master of the ordnance for Henry VII, and has six distinct components covering the period from 1495 to 1497. It includes an inventory of ordnance in the Tower of London in 1495, purchases of ordnance for the Scottish expedition, the shipping of guns to Scotland, the expenditure of equipment on both these expeditions and the remaining ordnance at the end of the accounts.[5] Clifford's accounts provide the most detailed set of information for artillery in the fifteenth century. They reveal not only the significant technological changes that had occurred between 1430 and the very end of the fifteenth century, but also the large quantities of equipment and personnel that were necessary to procure, operate and manage an artillery train by the latter date.

THE EXPEDITION OF 1430–2

The lifting of the siege of Orléans in 1429 led to a sequence of disasters for the English in France. The council of Henry VI sought to improve the deteriorating military situation on the Continent through the coronation of the young king in both England and France. The former was carried out on 6 November of the same year at Westminster Abbey, but the latter required a major military expedition to recover lost territories, including the coronation city of Reims.[6] The funding for this endeavour was agreed during the parliament of September 1429–February 1430. It was intended that an impressive force should accompany the king to France, with most of the indentures for the retinues being signed in mid-February. The participation of Henry VI resulted in the raising of an army of 4,792 men, making it the largest expedition sent to the Continent during his reign. The objectives of the campaign also meant that it was necessary to gather a large quantity of ordnance for sieges. On 24 February, John Hampton, master of the king's ordnance, was given a commission to assemble the artillery for the expedition. The terms of this patent specified the provision of artificers to construct carts and other equipment for the transportation of the king's ordnance, as well as the arrest of ships and sailors. Similar commissions were also granted to two other officials of the ordnance, John

4 TNA, E 36/8.

5 Note that the accounts of the shipping of ordnance to and from Scotland for the same campaign is also recorded in TNA, E 36/7, which has been printed by M. Oppenheim, ed., *Naval Accounts and Inventories of the Reign of Henry VII, 1485–8 and 1495–7, Publications of the Navy Records Society* (London: Naval Records Society, 1896).

6 Ralph A. Griffiths, *The Reign of King Henry VI: The Exercise of Royal Authority, 1422–1461* (London: Benn, 1981), pp. 189–90; Harris Nicolas, ed., *Proceedings and Ordinances of the Privy Council of England*, volume 2 (London: Record Commission, 1834–7), p. 223.

Louthe, clerk, and William Fleming, yeoman, in addition to the mason Robert Westerley. John Hampton's enrolled accounts detail the expenses incurred in the provision of artillery for the expedition which totalled £3,266 14½d.[7]

Assembling the Ordnance

The greatest expense was incurred in the purchase of equipment for the expedition which came to £2,072 4s. 8½d.[8] This amount was spent on a large variety of items, including guns, gunstones, carts, gunpowder, tools, bows, lances and other 'habiliments' of war throughout the period 1430–2. Twenty-three guns were acquired, made up of three large guns of iron, sixteen iron fowlers and four bronze fowlers. Significantly, the three large guns were named the *Henry*, *Hampton* and *Crown*, which symbolised the importance of these weapons to the campaign. The chance survival of the receipts for two of these guns, the *Henry* and *Crown*, means that it is possible to determine their provenance and prices. They were both purchased at Calais, the *Henry* of 6,780lbs from Godfrey Goykyn for £56 on 27 May, with the *Crown* of 7,122lbs acquired from Cons Beton for £59 7s. the next day. The identity of Cons Beton is obscure, but Godfrey Goykyn was a gunner who had been in English service since 1415 and appears to have been part of the Calais garrison.[9] It is also possible to determine where the gunpowder and saltpetre were purchased for the expedition. Hampton's accounts state that the gunners Giles Thorneton, Alex Mason and Godfrey Goykyn were sent by the king's command from Calais towards Bruges and Ghent in Flanders for the procurement of gunpowder and saltpetre for the war in France. It is probable that most, if not all, of the 1,204lbs of gunpowder and 5,164lbs of saltpetre purchased for the expedition was acquired from the Low Countries, as England was dependent on imports in this period.[10] A reference to the expenses incurred by John Louthe, clerk of the ordnance, in procuring guns from the city of London suggests that, as was the case with the 1428 expedition, the fowlers were purchased from merchants of the city. By contrast, it is not possible to determine where the rest of the ordnance was purchased from, as the accounts merely specify that they were obtained from 'diverse' locations in England and France.[11]

7 Anne Curry, 'The "Coronation Expedition" and Henry VI's Court in France, 1430 to 1432', in *The Lancastrian Court*, ed. Jenny Stratford (Donington: Shaun Tyas, 2003), pp. 29–52 (pp. 29–30); *CPR 1429–1436*, p. 44; TNA, E 364/69, rot. Q.

8 See Appendix F. TNA, E 364/69, rot. Q dorse.

9 TNA, E 101/52/22; E 364/69, rot. L dorse; E 101/45/5, m. 9.

10 David Cressy, *Saltpeter: The Mother of Gunpowder* (Oxford: Oxford University Press, 2013), pp. 40–2.

11 TNA, E 364/69, rot. Q dorse.

A large quantity of tools and goods were acquired for the operation and repair of the guns. This included items such as a pair of bellows, twenty chaldrons of sea coal, over 8,000lbs of worked and unworked iron, as well as 'crowes', staples and chains of iron, most likely for restoring damaged guns on the campaign. The artillery was also furnished with 5,200lbs of sulphur, 1,550 gunstones, 18 stocks for the fowlers and 105 pavises. Additional tools for the use of the ordnance company consisted of 996 hammers, 50 axes, 200 shovels, 200 spades and 42 'scoops', which would have been used in a variety of tasks such as cutting wood and digging trenches.[12] The accounts also provide information pertaining to the transportation of the artillery train. Four large carts were purchased for moving the guns and other equipment, with a further four long carts acquired specifically for use in the country of Picardy. These were to be conveyed by oxen and horses, for which eighty collars, twenty-five 'oxenbows', three lymour saddles, two collars of rope for cart horses and forty plates for moving oxen were provided. The numbers of animals that were used to move these carts is not stated. Some idea of the quantities necessary can be established, however, from a commission of 1420, which declared that sixty oxen and twenty-four horses were needed for moving four large carts which contained large guns.[13] A further £257 9s. 4d. was later given to Humphrey Stafford, earl of Stafford, by Hampton for the purchase of bows, arrows and lances on 8 January 1432 for military operations.[14] These purchases demonstrate that an artillery train required a large variety of equipment for the use of guns on campaign.

Hampton's accounts also record the payment of artificers for work on the ordnance and the freightage of the equipment to France, which came to £911 6s. 7d.[15] A total of ninety-five men were hired for the ordnance company to serve for six months from 1 May 1430. This consisted of twenty-six labourers, sixteen 'waymen', twelve gunners, twelve carpenters, twelve carters, six masons, six smiths, three purveyors and two clerks.[16] These workers were recruited by William Fleming and other unspecified persons, from the counties of Gloucestershire, Herefordshire, Worcestershire and Kent, undoubtedly as a result of the commission he had received in February.[17] The accounts furthermore state that diverse artificers were hired for 'making diverse necessary things' in parts of Normandy and France. The only location specifically mentioned in England is Oxford, where Walter Thomasson and five other gunners were paid to stay. It is unclear what tasks they carried out

12 For the definition of these terms see the Glossary.

13 *CPR 1416–1422*, p. 262.

14 TNA, E 364/69, rot. Q dorse.

15 TNA, E 364/69, rots. Q–Q dorse.

16 Later payments record that an ordnance company of between sixteen and twenty gunners, carpenters and mason served with Hampton in France, TNA, E 101/52/33, nos. 102, 111.

17 *CPR 1429–1436*, p. 44.

whilst located there, but Walter Thomasson had previously been involved in the construction of guns for the 1428 expedition, so it could have been for the same purpose. At least part of the equipment used for the expedition was assembled in London, with gunpowder, saltpetre, sulphur, charcoal and other items being stored in a hospital and house by Bishop's Gate. The ordnance was then freighted to France in six ships and one balinger.

Most of the guns provided for the expedition, however, were described as being received into the accounts from various individuals in Calais and Normandy.[18] It is unclear whether this artillery was financed from the resources of Lancastrian France or England, as no information exists in Hampton's accounts, or in other English Exchequer records, as to how it was paid for. Three large guns, the *Flouredelys* (Fleur-de-lis), *Newgate* and *Towerwharf of Cumbria*, were transferred by Richard Bokeland, treasurer of Calais, by indenture on 25 May 1430. The name of the *Flouredelys* suggests that it was constructed in France, whereas the other two were obtained from England. Further information is provided by Bokeland's accounts for Calais, which record that these guns were provided by Alex Mason and were test fired using 443lbs of gunpowder, 260lbs of sulphur and five gunstones.[19] Large quantities of guns were provided by John Clampard, smith of the king at Rouen, without indenture.[20] This consisted of a total of thirty-two stocked guns, made up of fourteen cart guns and eighteen fowlers. Clampard also transferred accessories for these guns, made up of chambers, chains, hooks and bolts, as well as smithy tools, such as two mandrels, one pair of tongs and one chisel. Additional equipment for the cart guns was provided by two other men from the same city. William Hampton, master carpenter of the king at Rouen, gave fourteen stocks and sixteen stakes of wood, whereas the vicomte of Rouen contributed sixteen carts.

Further equipment was handed over by William Appulby, master of the ordnance of the king in Normandy. This included one gun called *Coloffre*, one fowler, 1,672lbs of sulphur and 717lbs of gunpowder. Large quantities of gunpowder, saltpetre and sulphur were also obtained from London, Rouen and Harfleur. In the case of the latter a set of instructions exists as to how this material was obtained. On 16 April 1430, a warrant under the privy seal was addressed to William Minors, captain of Harfleur, to deliver to William Hampton all guns and ordnance in his keeping, except those needed for the town's defence.[21] In the event it appears that relatively little ordnance was still stored in Harfleur as only 3,597lbs

18 See Appendix G.

19 TNA, E 364/65, rots. D–D dorse.

20 Note that this reference and the following do not provide dates for when equipment was transferred.

21 John Bree, *The Cursory Sketch of the State of the Naval, Military, and Civil Establishment, Legislative, Judicial, and Domestic Economy of this Kingdom during the Fourteenth Century;*

of saltpetre was handed over by the captain and other officers of the garrison. A further 7,925lbs of gunpowder, 4,438lbs of saltpetre and 4,648lbs of sulphur was received by Hampton at London and Rouen from William Warner, a gunpowder maker at London, and Robert Westley.

Use of the Artillery on Campaign

Henry VI and his army were transported to Calais in the spring of 1430, with the king arriving at the town on 23 April.[22] The dangerous military situation in France, however, meant that he was not to be crowned until December 1431 and in Paris as opposed to Reims. French victories since the previous year had significantly reduced the territory under English control, particularly in the Île-de-France. It was therefore necessary to secure the area around Paris, as well as to clear a path to Reims for the intended coronation. Reims had previously been under the control of England's main ally, Philip, duke of Burgundy, whose assistance was sought for the forthcoming campaign.[23] The objectives of the expedition, therefore, principally consisted of the capture of key towns and cities in north-eastern France. This meant that the military operations of 1430–2 were characterised by siege warfare, in which artillery played an important role. The English were well equipped in this regard, due to the concerted effort made to acquire a large number of guns from England, Normandy and Calais. Additional support was provided by the Burgundians who themselves possessed a large quantity of ordnance.[24]

The subsequent military campaign can be partly reconstructed from the places listed in Hampton's accounts and from narrative sources such as the chronicles of Enguerrand de Monstrelet.[25] A number of locations in the Île-de-France, Picardy and Normandy are mentioned by the former, including Compiègne, Torcy, Brie-Comte-Robert, Château Gaillard, Clermont and Louviers. These are listed in connection with the use of equipment by soldiers in various offensives during the period of the accounts. Unfortunately, a breakdown of how the ordnance was expended is not provided, but it seems reasonable to assume that the guns sent

with a Particular Account of the Campaign of King Edward the Third, in Normandy and France, in the Years 1345 and 1346 (London: Printed for the Author, 1791), p. 149.

22 Griffiths, *The Reign of King Henry VI*, p. 191; KHLC, NR/FAC2, f. 110r; Curry, 'The "Coronation Expedition" and Henry VI's Court in France', p. 31.

23 For payments for military assistance see Richard Vaughan, *Philip the Good: The Apogee of Burgundy* (London: Longmans, Green and Co. Ltd, 1970), p. 17.

24 For Burgundian artillery see Robert D. Smith and Kelly DeVries, *The Artillery of the Dukes of Burgundy 1363–1477* (Woodbridge: Boydell Press, 2005).

25 L. Doüet-d'Arcq, ed., *La Chronique d'Enguerran de Monstrelet, en Deux Livres avec Pièces Justificatives, 1400–1444*, volume 4 (Paris: Société de l'Histoire de France, 1860).

from England were used in these offensive actions alongside others assembled from other sources in France. It is likely that Hampton took possession of artillery constructed in France during the duration of the expedition, from Calais in April and then later from Rouen when the king travelled there in July.[26] From the beginning it appears that weaponry was distributed by Hampton for different armies operating in northern France. Multiple places were under siege by English forces at the start of the year, including the castles of Torcy and Château Gaillard.[27] The arrival of the army and artillery from England meant that multiple operations could be undertaken by the Anglo-Burgundians from late spring onwards.

In May the town of Compiègne was besieged by the army of the duke of Burgundy, supported by an English force.[28] At about the same time, John Mowbray, duke of Norfolk, was said to have subjected many fortresses in the Île-de-France to the obedience of King Henry including Dammartin and others. A separate force led by the earl of Stafford captured the town and castle of Brie-Comte-Robert.[29] The Anglo-Burgundians, however, experienced setbacks at the end of 1430. A concerted effort had been made by the army besieging Compiègne to take the town, but despite using large numbers of guns, the siege had to be abandoned by November.[30] Similarly, an English force made up of many of the soldiers sent over in April was forced to lift the siege of Louviers at around the same time. The reversal at Louviers was particularly unfortunate as it meant that the route to Paris was too unsafe for Henry VI to visit. The decision was therefore taken to send another army of 3,450 men from England in early 1431.[31] There is no evidence that any additional ordnance was sent with it to France at this time.

Further campaigning was undertaken by Anglo-Burgundian armies in the Île-de-France and Normandy in 1431. A contingent was sent by Bedford and the king's council at Paris to attack castles on the borders of the Île-de-France. According to Enguerrand de Monstrelet, they were said to number 1,200 soldiers accompanied by a plentiful number of carts and carriages with guns, artillery and other instruments of war. After capturing multiple castles including Gournay, they

26 Griffiths, *The Reign of King Henry VI*, p. 191.

27 The siege of Torcy had begun in 1429 and the siege of Château Gaillard by April 1430, Joseph Stevenson, ed., *Letters and Papers Illustrative of the Wars of the English in France during the Reign of Henry the Sixth, King of England*, volume 2, part 2 (London: Longman, Green, Longman, and Roberts, 1864), p. 539; Doüet-d'Arcq, *La Chronique d'Enguerran de Monstrelet*, volume 4, pp. 368–70.

28 Smith and DeVries, *The Artillery of the Dukes of Burgundy 1363–1477*, p. 100; Doüet-d'Arcq, *La Chronique d'Enguerran de Monstrelet*, volume 4, pp. 390–1.

29 Doüet-d'Arcq, *La Chronique d'Enguerran de Monstrelet*, volume 4, p. 405.

30 Ibid., pp. 396–8, 402–5, 409–20; Smith and DeVries, *The Artillery of the Dukes of Burgundy 1363–1477*, pp. 102–3.

31 Curry, 'The "Coronation Expedition" and Henry VI's Court in France', p. 42; Griffiths, *The Reign of King Henry VI*, p. 191.

marched on Lagny-sur-Marne. One of the English commanders, John Fitzalan, earl of Arundel, was subsequently able to use a large bombard to destroy the arc of a drawbridge linking a bulwark to the town; it was then successfully stormed. After this, the army tried to capture the town through assaults and eight days of bombardment but were forced to retreat to Paris having suffered heavy casualties.[32] The English had more success in Normandy at the siege of Louviers which had been resumed in spring of the same year. Two masters of the ordnance served with the besieging army, William Appulby, master of the ordnance of the king in Normandy, and Philebert de Moulins, master and surveyor of the king's artillery in France. Thomas Blount, treasurer of Normandy, was directed to send additional equipment for the siege in June; this consisted of thirty mantlets and twenty carts to transport guns to the town, with gunstones being sent later, in August. Louviers was eventually captured in October of the same year.[33] These victories meant that the path to Paris was secure enough for Henry VI to be safely escorted to the city for his coronation in December.[34]

These campaigns, therefore, made extensive use of artillery, which is reflected in the list of ordnance expended or depleted from Hampton's accounts.[35] This included a third of the sixty guns procured for the expedition, made up of twenty fowlers, together with large quantities of gunpowder, saltpetre, sulphur, gunstones, iron and other equipment. It is likely that much of this equipment was used up in the conduct of military operations, but losses through capture, theft and accidents are also possibilities. The expenditure of two mandrels and over 6,000lbs of iron work suggest that repairs were carried out to the equipment during the expedition, most likely for iron guns, gun chambers, stocks and carts. Therefore it is possible that a greater proportion of the iron guns were damaged in siege warfare than is indicated by these accounts. In particular, major efforts

32 L. Doüet-d'Arcq, ed., *La Chronique d'Enguerran de Monstrelet, en Deux Livres avec Pièces Justificatives, 1400–1444*, volume 5 (Paris: Société de l'Histoire de France, 1861), pp. 27–30.

33 Carole Rawcliffe, 'Blount, Sir Walter (d. 1403)', *Oxford Dictionary of National Biography, Oxford University Press, 2004* http://www.oxforddnb.com/view/article/2699 [accessed 7 April 2015]; Charles Purton Cooper, ed., *Appendices to a Report on Rymer's Fœder, Appendix D* (Public Record Office, 1869), pp. 393, 395–6; Andy King, 'Gunners, Aides and Archers: The Personnel of the English Ordnance Companies in Normandy in the Fifteenth Century', *Journal of Medieval Military History*, 9 (2011), 65–75 (p. 66); Curry, 'The "Coronation Expedition" and Henry VI's Court in France', pp. 43–4; Anne Curry, ed., *The Parliament Rolls of Medieval England, 1275–1504: X Henry VI. 1422–1431*, pp. 435–7.

34 Curry, 'The "Coronation Expedition" and Henry VI's Court in France', p. 31; Griffiths, *The Reign of King Henry VI*, p. 193.

35 See Appendix H. Unfortunately no dates are provided which means that it is not possible to determine where and how the equipment was damaged or used up.

would have been made to maintain the prestigious and expensive bombards. It is also of interest that a much higher proportion of the gunpowder, saltpetre and sulphur was expended than of the gunstones. This may have been a reflection of the widespread availability of gunstones in Normandy or their potential for being reused after firing.[36]

Finally, at the end of the expedition in 1432, the remaining ordnance was distributed to different officials of the king in Lancastrian France, instead of being returned to England. Most of the equipment was given to William Appulby, master of the ordnance of the King in Normandy by four indentures in Rouen. It included the seven large named guns, *Henry*, *Crown*, *Hampton*, *Coloffre*, *Flouredelys*, *Newgate* and *Towerwharf of Cumbria*, along with all fourteen of the cart guns and seven fowlers; in addition to iron, tools, gunpowder and harnesses for animals. These guns would have therefore made up an important part of the artillery available to Appulby for future operations. A smaller quantity of similar items had also been transferred to Appulby's predecessor, William Gloucester. The rest of the ordnance was allocated for the defence of three strategic locations in Normandy in December 1431. The towns and castles of Caen, Gournay and Touques were each allocated four fowlers, 660lbs of gunpowder and other equipment.

THE EXPEDITIONS OF 1497

Relations between England and Scotland deteriorated in 1495 due to the support given to the pretender, Perkin Warbeck, by James IV, King of Scots. Earlier the same year, Warbeck had attempted to invade England with an army funded by Maximilian I, King of the Romans, and Margaret of Burgundy. After the failure of the landing in Kent, Warbeck was forced to flee to Ireland and then later to Scotland. He was well received by James IV who acknowledged him as Richard, duke of York, and arranged a marriage to his distant cousin, Katherine Gordon. This led to the Scottish decision to go to war with England in the summer of 1496.[37] In September, a Scottish army launched a raid into Northumberland and destroyed some peel towers, but was forced to withdraw after only two days. Henry VII responded to this attack by holding a council at Sheen in October, where the decision was made to invade Scotland with two large armies and a fleet. At a Great Council, held between 24 October and 6 November, it was decided that a grant of £120,000 would be made to the king for financing the war. This subsidy was subsequently confirmed by Parliament in January 1497, although the scale

36 For quarries in Caen see J. L Kirby, ed., *Calendar of Signet letters of Henry IV and Henry V (1399–1422)* (London: H. M. S. O., 1978).

37 Ian Arthurson, *The Perkin Warbeck Conspiracy, 1491–1499* (Stroud: Sutton, 1997), pp. 106, 112–16, 121, 133.

of taxation caused deep resentment.[38] Responsibility for supplying the artillery necessary for such a substantial undertaking was given to Sir Robert Clifford, master of the ordnance.

The resources available to Clifford at the end of 1496 can be ascertained from his accounts, which include an inventory of ordnance in the Tower of London in 1495, compiled after his appointment to the office of master of the ordnance.[39] Much of this equipment was artillery left over from the campaigns of 1489–92 in Brittany and France.[40] This inventory is particularly useful as it provides insights into the significant changes that had occurred to artillery between 1430 and 1497. This can be seen from the diverse types of guns recorded, comprising of ninety-four bombardelles, great curtows, demi-curtows, serpentines, falcons, small guns, pot guns, murderers, other guns and organ guns, as well as nineteen hackbuts. The presence of a variety of large, medium and small guns, reflected the increasing specialisation in how artillery was used at the end of the fifteenth century, both on the battlefield and in siege warfare. A range of ammunition was available for these weapons, with hundreds of iron shot and gunstones for the larger ordnance and lead shot for the smaller ordnance. The stores also contained items for the horses of the ordnance, such as over 611 horse harnesses, 5,060 horse shoes, 190 horse collars, along with carts, saddles and wheels. Other equipment included large quantities of gunpowder, made up of seven lasts and ten barrels, as well as rammers, charging ladles, moulds and tampons. At least part of the artillery was intended for a defensive purpose, with three guns located in the bulwark, two in the 'grene' gate and two over the 'capstan brege'.[41] Nevertheless, the bulk of the ordnance used for the Scottish expedition was already located in the arsenal of the Tower of London. Other equipment was also received from Kenilworth by Clifford at this time, which included twelve falcons, twenty-three half barrels of gunpowder and eleven hackbuts.[42]

Assembling the Ordnance

The poor state of the equipment meant that a good deal of refurbishment was necessary. This can be seen from a reference in Clifford's accounts to repairs carried out by horse harness makers for harnesses that were 'hurt and perished

38 Ibid., pp. 149, 153–4; Sean Cunningham, 'National War and Dynastic Politics: Henry VII's Capacity to Wage War in the Scottish Campaigns of 1496–7', in *England and Scotland at War, c.1296–c.1513*, ed. Andy King and David Simpkin (Leiden: Brill, 2012), pp. 297–328 (pp. 306–7).

39 TNA, E 36/8, ff. 90r–95r. See Appendix C.

40 TNA, E 36/15, f. 33v

41 The latter appears to be a pulley bridge of some description, TNA, E 36/8, f. 90v.

42 Ibid., f. 95r.

by rats and other vermin'.[43] Clifford, therefore, between December 1496 and May 1497, paid a large number of workers to refurbish the artillery stored in the Tower of London. This progress was initiated by a warrant given on 1 December 1496, which authorised the transfer of £622 19s. 8d. in part payment of £1,000 for the procurement of ordnance.[44] Clifford benefited from being able to draw upon an established body of royal gunners for the expedition to Scotland. Since 1494 a small number of gunners had been in receipt of annual wages from the Exchequer (see Table 2). These men came from a variety of backgrounds, such as the gunfounder William Newport, who had supplied guns for the king during the Brittany campaigns of 1489–91, with others originating from Continental Europe, such as Blake and Symond Ballard who were Swiss.[45] In 1497 their numbers were temporarily augmented by the recruitment of large numbers of gunners from the Low Countries. The individual tasked with this mission seems to have been Philip Lokier, who was sent to Flanders in December 1496, where he purchased handguns and hired seven master gunners. The clerk of the ordnance, William Furkins, was eventually responsible for paying the wages of 1,000 men for the ordnance, of whom 200 were gunners.[46]

The refurbishment of the artillery for the expedition began in December 1496, when multiple commissions were given out for the recruitment of harness makers, smiths, wheelwrights and other artificers.[47] Over the following five months, gunners and other professionals, based mainly at the Tower of London, were paid to work on the ordnance. On 22 December an advance payment of £10 was given to the Southwark goldsmith, Henry Ftner, to supply iron from Ashdown Forest in Sussex for the works of the king.[48] In January and February, work was carried out in the Tower of London by eight horse harness makers on the repair and construction of horse harnesses and collars. Three of these men, John Vaunce, John Smyth and Thomas March, were also paid for supplying seventy-four carriages for great ordnance and thirty carriages for the smaller ordnance, in addition to horse

43 Ibid., f. 29r.

44 Note that large quantities of other equipment was purchased for the ordnance including bows, arrows and bills. TNA, E 36/8, f. 83v; E 405/79, m. 15d.

45 Blake was described as being Swiss in 1509, J. S. Brewer, 'Henry VIII: December 1509', in *Letters and Papers, Foreign and Domestic, Henry VIII, Volume 1, 1509–1514* (London, 1920), pp. 127–44, http://www.british-history.ac.uk/letters-papers-hen8/vol1/pp127-144 [Accessed 29 January 2015]; TNA, E 405/78, m. 39d; E 36/124, f. 82v.

46 TNA, E 405/79, mm. 15r, 35d; Ian Arthurson, 'The King's Voyage into Scotland: The War that Never Was', in *England in the Fifteenth Century: Proceedings of the 1986 Harlaxton Symposium*, ed. Daniel Williams (Woodbridge: Boydell Press, 1987), pp. 1–22 (p. 8).

47 *CPR 1494–1509*, pp. 89–91.

48 TNA, E 36/8, f. 25r.

Table 2 Royal gunners in receipt of annual wages 1494–1500

Names	Dates
Robert Anylety[a]	1498–9
Blake Ballard[b]	1494–9
Symond Ballard[c]	1494–9
Thomas Bowyer[d]	1495–6
Henry Bsecrant[e]	1498–9
Antony de Capetto[f]	1499–1500
Gerard Farlrad[g]	1499–1500
Richard Faucon[h]	1494–9
Robert Fisher[i]	1497–9
Thomas Greves[j]	1499–1500
Nicholas van Herlond[k]	1498–9
John van Hoven[l]	1499–1500
Thomas Isott[m]	1494–7
William Ive[n]	1494–7
William Kere[o]	1497–8
Jacob Kyrkeby[p]	1498–1500
Henry Martin[q]	1496–8
William Newport[r]	1494–9
Rowland Peys[s]	1494–7
Peter Rore[t]	1497–8

Sources: TNA – [a] E 405/80, m. 10d; [b] E 405/79, mm. 8d, 10d, 22r, 38r; [c] E 405/79, mm. 8d, 22r, 38r; [d] E 405/79, m. 10d; [e] E 405/80, m. 10d; [f] E 405/80, mm. 20r, 36d, 47v; [g] E 405/80, m. 20r; [h] E 405/79, mm. 8d, 10d, 22r; E 405/80, mm. 10d, 34d, 36d, 39d; [I] E 405/79, m. 38r; E 405/80, mm. 10d, 34d, 36d, 39d; [j] E 405/80, mm. 20r, 36d, 39d; [k] E 405/80, m. 10d;[l] E 405/80, m. 20r; [m] E 405/79, m. 8d; [n] E 405/79, mm. 8d, 10d, 22r; [o] E 405/79, m. 38r; [p] E 405/80, mm. 10d, 20r; [q] E 405/79, mm. 8d, 10d, 22r, 38r; [r] E 405/79, mm. 8d, 10d, 22r, 38r; E 405/80, mm. 10d, 20r, 34d, 39d, 47r; [s] E 405/79, mm. 8d, 10d; [t] E 405/79, m. 38r.

draughts.[49] Richard Faucon, elsewhere described as the king's master gunner, also supplied fourteen barrels of gunpowder at around this time.[50] Cast iron shot from Ashdown Forest was later provided by Symond Ballard in April, with guns being tested at Mile End in the same month.[51]

The approaching deadline for the shipment of the ordnance to Scotland, by the middle of May, meant that an increase in activity at the Tower of London occurred at this time, with at least eight wheelwrights being employed for the sorting and binding of carts from 11 April until 12 May, in addition to fourteen gunners who cast lead shot at the same time.[52] Numerous items of ironwork were forged and constructed by the smith, John Trewchard, including bolts, staples, nails, forelocks, hooks, bands and bolters, for stocks, cranes, wheels and other items. William Newport was also paid for constructing nine falcons and four serpentines of bronze.[53] These preparations seem to have been completed by Ascension Day (14 May), when labourers were paid to convey hackbuts from the House of the Ordnance to the walls of the White Tower, where they were fired in the presence of the king, possibly as a way of announcing the start of the campaign.[54]

Calais was also an important location for supplying ordnance for the expedition, as had been the case in 1430. It is difficult to determine exactly how much equipment was sent from the territory, due to the lack of surviving information for Calais in the reign of Henry VII, but references in Clifford's accounts suggest that large quantities of supplies were derived from there.[55] On 17 March, the carter John Horgan was paid for the carriage of twenty loads of 'stuff' sent from Calais to the Tower. It is unclear what this particular item refers to, but later entries record the receipt of gunpowder and repairs to carts that came from the same location.[56] Other equipment appears to have been sent from Calais directly to the north. This included at least two guns dating from the reign of Edward IV, the *Great Edward* and *Little Edward*, for which horse harnesses and draughts were ordered in January.[57]

The ordnance was then subsequently transported by twenty-three ships, whose masters had indented with Clifford on 17 May. The latter's accounts show that a total of 49 guns and 181 hackbuts were freighted to Berwick-upon-Tweed

49 Ibid., ff. 29r, 30v, 32v, 39r.

50 Ibid., ff. 24r, 30v, 33r.

51 Ibid., ff. 38r, 49r, 51v, 65r, 71r.

52 Arthurson, 'The King's Voyage into Scotland', p. 14; TNA, E 36/8, ff. 66r–66v, 67v, 72v.

53 TNA, E 36/8, ff. 66v, 68v–69v, 70v, 76r.

54 Ibid., f. 78r.

55 Grummitt, *The Calais Garrison*, pp. 119, 192.

56 TNA, E 36/8, ff. 39v, 72r–72v.

57 Note that it is possible that these guns may have been left behind in Newcastle after having been transported to the north for the 1482 expedition to Scotland, as they are not recorded in the Calais inventories of 1485 or 1488, TNA, E 364/119/36, rots. C, D.

and Newcastle, together with large quantities of other equipment.[58] The former comprised two bombardelles, called *Portcolies* and *Rose*, two curtows, called *Tower* and *Dragon*, eleven demi-curtows, called *Carlisle, Dover, Berwick, Leopard, Rysbank, Calais, Pontefract, Windsor, Gusynes, Bamburgh* and *Hammes*, six serpentines, of which one was called *Greyhound*, and twenty-eight falcons. Interestingly, these guns appear to have been constructed in a standardised fashion firing shot and gunpowder according to their class. The bombardelles are described as firing iron shot weighing 225lbs using 50lbs of gunpowder, the curtows as firing iron shot weighing 120lbs using 60lbs of gunpowder, the demi-curtows as firing iron shot weighing 40lbs using 20lbs of gunpowder, the serpentines as firing iron shot weighing 20lbs using 20lbs of gunpowder and the falcons as firing lead shot weighing 1lbs using 1lbs of gunpowder. This can also be seen from the equipment provided with the guns. For instance, each of the eleven demi-curtows was listed as possessing 150 iron shot, 50 gunstones, 2 charging ladles, 2 rammers and 20 horse harnesses.[59] Other equipment shipped to Scotland included large quantities of gunpowder, tampons, lead shot and iron dice, as well as carts, wheels and tools.

Blackheath Campaign

Henry VII's carefully planned invasion of Scotland was suddenly undone, however, by a major rebellion which broke out in Cornwall on 14 May 1497. Widespread unhappiness over the heavy taxes imposed for the war with Scotland meant that an initial rising in West Cornwall led by the Cornish blacksmith An Gof rapidly spread eastwards. The rebels received further support after they left Cornwall, particularly in Somerset, but were unable to obtain control of major cities, like Exeter and Bristol.[60] The rebellion put the king in a difficult position. Two-thirds of the army raised for the Scottish campaign was on the way or already in the north, with only the final third uncommitted by the beginning of June.[61] Henry's response was to entrust the King's Chamberlain, Giles, Lord Daubeney, with blocking the rebels' path to London with the retinues that were available, whilst the king moved to Woodstock to muster an army.[62]

58 See Table 3. Note that there is a discrepancy between TNA, E 36/8 and E 36/7 as to the number of hackbuts, 100 and 181, in this instance I have gone with the latter figure.

59 Note that there is slightly more variation with the other guns, e.g. one bombardelle had 65 iron shot and 30 gunstones, whereas the other had 67 iron shot and 36 gunstones.

60 In the case of the former an agreement was reached whereby the rebel army was kept out but the commanders were allowed to march through the city. Arthurson, *The Perkin Warbeck Conspiracy*, pp. 163–5.

61 Arthurson, 'The King's Voyage into Scotland', p. 11.

62 Arthurson, *The Perkin Warbeck Conspiracy*, p. 165.

TABLE 3 Summary of ordnance shipped to Berwick and Newcastle in 1497.

Item	Quantity
Bombardelles	2
Curtows	2
Demi-curtows	11
Serpentines	6
Falcons	28
Hackbuts	181
Iron shot	3,320
Gunstones	1,554
Horse harnesses	375
Gunpowder	3 lasts, 18 barrels, 2 firkins
Lead shot	28 baskets, 188 lead shot
Iron dice	9 baskets, 1196 iron dice
'Fare' carts	13
Close carts	8
Tampons	12,100

Source: TNA, E 36/8, ff. 1r–7r.

Guns for use on the anticipated battlefield were also obtained from the Tower of London for equipping the royal army. This can be seen from an unusually detailed set of payments which records the transportation of artillery by land prior to the battle of Blackheath on 17 June.[63] The first part lists the movement of nine whole and eight 'half' carriages of ordnance, in stages, from the Tower of London to Woodstock by diverse carters. The equipment moved to the latter would have comprised a variety of items and weapons, including bows, bills and gunpowder, but it also included falcons, which due to their small size were suitable for use on the battlefield. This can be seen from the second part of the payments, which records the movement of eight falcons from Woodstock, where the king was located, to join Daubeney's army at Hounslow Heath, where the carters were paid for a further two days to wait upon the guns, before the falcons were finally conveyed to Blackheath. Responsibility for guarding the ordnance whilst it was being moved was entrusted to the carters, whereas at Hounslow Heath additional

63 See Appendix I. The entries do not provide dates but it is likely to have moved after 6 June when the ordnance company was assembled.

night watchmen were hired.[64] These entries also reveal interesting information about the wear and tear upon equipment, which occurred during the transportation of ordnance over a total distance of more than 142 miles by land. Repairs were carried out to 'close' and gunpowder carts, mainly to replace damaged axles, shafts and wheels, which were also greased with tallow to prevent further deterioration.

The ordnance company for the campaign was assembled by 6 June when diverse artificers and gunners were recorded as being retained in the king's wages for a month 'attending upon the ordnance'.[65] A total of 236 men were recruited, made up of the yeoman of the ordnance, his 2 assistants, 2 'piyone(r)s', 20 bowyers, 1 stringer, 12 smiths, 6 fletchers, 6 horse harness makers, 7 'gardtramells', 7 wheelwrights, 1 surgeon, 49 gunners and 122 pioneers.[66] This included almost all of the gunners in receipt of royal wages at this time, many of whom had previously been paid to cast lead shot and work on other equipment in preparation for the Scottish expedition.[67] The size of the ordnance company in this campaign demonstrates that a large number of professionals were necessary to operate and maintain an artillery train by 1497. Surprisingly, however, it appears that the artillery transported to Blackheath played little or no part in determining the outcome of the battle. Clifford's accounts record that a number of gun accessories, such as bags of leather, charging ladles, axles or horse harnesses, were spent or lost at Blackheath, which suggests that the falcons were prepared for action. By contrast, however, no gunpowder or ammunition was listed as having been expended; therefore the guns are unlikely to have been fired. Instead this was a battle which was fought and won using traditional weapons, such as bows and bills, sizeable quantities of which are recorded as having been spent at Blackheath.[68] The reason for this may have been because it was difficult to effectively use field artillery offensively on the battlefield in this period, as the royal army advanced to attack the Cornish, or because it was simply unnecessary as the rebel forces were quickly overwhelmed.

64 TNA, E 36/8, ff. 76v–77r.

65 Ibid., ff. 78v–82r.

66 'Pioneers' appear to have been labourers tasked with moving guns and digging trenches: see C. T. Onions, ed., *The Shorter Oxford English Dictionary*, volume 2 (Oxford: Clarendon Press, 1973), p. 1,735. Whereas the term 'gardtramelles' seems to refer to stakes used on the battlefield: see Oppenheim, *Naval Accounts and Inventories of the Reign of Henry VII*, p. 121.

67 TNA, E 36/8, ff. 58v, 66r–66v, 67v, 72r.

68 For example, 300 sheaves of arrows are listed as being expended at the battle. Ibid., f. 96v.

Scottish Campaign

James IV made little attempt to take advantage of the chaos caused by the Western Rising, with only a couple of minor raids launched into England in June.[69] This was perhaps due in part to the large numbers of English soldiers which had already been sent to the north. Nevertheless, at the beginning of August, James did cross the border with his army and besieged Norham Castle in Northumberland. The Scottish artillery bombarded the castle to little effect and the siege had to be abandoned after ten days. This was due at least in part to the arrival of the army of Thomas Howard, earl of Surrey, in Berwick via Newcastle at around the same time.[70] Surrey was therefore able to collect the ordnance which had been sent to those places earlier in the year. In addition to this, he also obtained one curtow and two demi-curtows from the castle at Newcastle, as well as one serpentine and five serpentines provided by Robert, Lord Willoughby, commander of the English fleet at Berwick.[71] This meant that a total of fifty-eight guns of a variety of sizes and 100 hackbuts were available to the English army. Only twenty-seven guns, consisting of one curtow, two demi-curtows and twenty-four falcons, in addition to fifty hackbuts, were transported into Scotland with the English army on 16 August, with most of the large guns being left behind at Berwick.[72]

The reason for this decision appears to have been due to the limited objectives of the campaign in August, which was focused upon countering the Scottish army in the field as opposed to attacking major settlements. On 16 August, the English attacked and demolished the peel towers of 'Cawmyllis', 'Hoten Hall', Edington and Foulden before proceeding on to the Tower of Ayton the following day, which surrendered after a bombardment of three hours and was then destroyed using gunpowder.[73] A total of twelve hackbuts and five falcons broke under the strain of fire in these actions, with the latter described as being in twenty-eight pieces. The damage caused to the smaller guns is unsurprising considering that 1,110 lead shot were fired by the falcons and hackbuts; by contrast, only one iron shot was expended by a curtow, twenty-four iron shot by the demi-curtows and fifty lead shot by the serpentines.[74] On the following day a stand-off occurred between the English and Scottish armies which were drawn up to give battle. The English guns

69 Arthurson, *The Perkin Warbeck Conspiracy*, p. 169.

70 Ibid., pp. 172–3; Cunningham, 'National War and Dynastic Politics', p. 310.

71 Oppenheim, *Naval Accounts and Inventories of the Reign of Henry VII*, p. 103.

72 TNA, E 36/8, f. 9r.

73 I have been unable to establish the modern names for all of these places.

74 Note that no serpentines are recorded as having been sent into Scotland on 16 June, so it is possible that additional ones were sent on afterwards. TNA, E 36/8, f. 9r; Oppenheim, *Naval Accounts and Inventories of the Reign of Henry VII*, pp. 103, 129–31.

were prepared for action but were unused as neither side initiated hostilities, after which both sides withdrew.[75]

At the end of the campaign most of the ordnance was shipped back to the Tower of London (see Table 4). This included thirty-six of the remaining fifty-four guns, made up of two bombardelles, three curtows, ten demi-curtows, five serpentines and sixteen falcons; as well as 144 hackbuts. Other items that were returned after the expedition finished included the bulk of the ammunition, gunpowder and horse harnesses. Some of the equipment was lost whilst in the process of being transported by sea, including twenty-four hackbuts, one barrel of gunpowder and one and a half barrels of saltpetre.[76] The rest of the ordnance was left in the custody of William Pawne in Berwick, consisting of one curtow, two demi-curtows, four serpentines and eleven falcons, as well as gunpowder, ammunition and horse harnesses (see Table 5).

TABLE 4 Summary of ordnance returned to the Tower of London in 1497.[77]

Item	Quantity
Bronze bombardelles	2
Curtows of bronze	3
Demi-curtows of bronze	10
Serpentines of bronze	4
Serpentines of iron	1
Falcons of bronze	16
Hackbuts of iron	144
Rammers of 'tre' for great guns	21
Charging ladles	78
Gunpowder	4 lasts, 1 barrel, 1 firkin
Iron shot	3,526
Gunstones	1,633
Lead shot	30 baskets
Dice of iron for pellets	41 baskets
Shipping crane with her raising 'gynne'	1

75 TNA, E 36/8, f. 9r; Oppenheim, *Naval Accounts and Inventories of the Reign of Henry VII*, pp. 129–31; Cunningham, 'National War and Dynastic Politics', p. 310.

76 Oppenheim, *Naval Accounts and Inventories of the Reign of Henry VII*, p. 131.

77 Ibid., pp. 121–4.

Item	Quantity
Loading 'gynnes'	2
Moulds for pellets	4 baskets
Tampons	12,100
Close carts for gunpowder	7
Carts with 'tillettes' for gunshot	8
Horse harness for bombardelles	25
Horse harnesses for curtows and demi-curtows	134
Horse harnesses for serpentines, falcons and carts	350

Source: Oppenheim, *Naval Accounts and Inventories of the Reign of Henry VII*, pp. 121–4.

TABLE 5 Summary of ordnance left in Berwick with William Pawne in 1497.

Item	Quantity
Curtows of bronze	1
Demi-curtows	2
Serpentines of bronze	2
Small serpentines of bronze	2
Falcons of bronze	11
Iron shot for curtows	125
Iron shot for demi-curtows	102
Gunstones for demi-curtows	20
Lead shot	4 baskets
Cart horse harnesses for curtows	25
Horse harnesses for demi-curtows	38
Horse harnesses for serpentines	12
Horse harnesses for falcons	36
Gunpowder	17 barrels

Source: Oppenheim, *Naval Accounts and Inventories of the Reign of Henry VII*, p. 127.

CONCLUSION

This comparison has shown that the artillery used in expeditions changed significantly between 1430 and 1497. Many new types of guns were in use by the end of the century, which demonstrates how artillery was becoming increasingly specialised for use on the battlefield and in siege warfare. The weaponry deployed in the 1497 expedition also differed markedly from what was used in the French and Scottish expeditions of 1475 and 1481–2, indicating that comparatively rapid technological change occurred in the last quarter of the fifteenth century. Similarly, notable shifts can be detected in the preferred form of metal employed in gun-making and in the types of ammunition that was provided for artillery. For much of the fifteenth century, wrought iron was used to make most English guns, yet by the 1490s, bronze guns predominated in expeditions to France and Scotland. Different types of ammunition also came into use towards the end of the century, including iron dice and cast iron shot, which suggests that new developments in metallurgy, such as the casting of iron in Ashdown Forest in East Sussex, were the necessary requisites for these changes.[78] Furthermore, the large quantities of ammunition provided for the 1497 expedition also indicates that guns were intended to be used, and were actually used more frequently, as can be seen from the heavy expenditure of shot during the attack on the Tower of Ayton.

The transportation of artillery also changed significantly throughout the course of the fifteenth century. John Hampton's accounts of 1430–2 list 300 yokes for oxen and 79 horse harnesses, by contrast, 375 horse harnesses were shipped to Scotland in 1497. Similar numbers of guns were used in both expeditions, 60 in 1430–2 and 58 in 1497, but this comparison reveals that the equipment was moved differently. This change is likely to have occurred because although oxen were stronger and had greater endurance than horses, they were also slower and harder to manoeuvre. The shift to using horses, therefore, made it easier to transport artillery on campaign, particularly the smaller guns for use on the battlefield. This can also be seen with the carriages provided for the movement of ordnance. In 1497, all of the guns that were shipped to Scotland were described as being individually stocked and carted upon two or four wheels, which meant that they had their own gun carriages. The evidence for the movement of falcons during the Blackheath campaign suggests that they were moved by carters using carthorses, whereas it appears that the guns in 1430–2 lacked their own individual carriages, as they were loaded on to large carts pulled mostly by oxen and then had to be unloaded when they reached their destination. Artillery was therefore much more mobile by the end of century. It is likely that this change occurred at some point between the 1450s and 1470s,

78 Brian Awty and Christopher Whittick, 'The Lordship of Canterbury, Iron-Founding at Buxted, and the Continental Antecedents of Cannon-Founding in the Weald', *Sussex Archaeological Collections*, 140 (2002), pp. 71–81.

with the development of field artillery. The provision of 'fare' and 'close' carts for the movement of ordnance provides further evidence that transportation had become more specialised and sophisticated.

It is also apparent that the production of artillery became more centralised throughout the course of the fifteenth century. In 1430, as was also the case in 1428, guns were supplied by gunners and merchants from a variety of locations in England and Lancastrian France. By contrast, the activities associated with the production and refurbishment of equipment in 1497 was focused upon the Tower of London. This change was partly due to strategic and political factors, such as the loss of Normandy and Henry VII's distrust of the Calais establishment.[79] It was also as a result of changes to the administration of the ordnance and the growing professionalisation of the personnel associated with the royal artillery. John Hampton was appointed solely for the duration of the 1430–2 expedition, whereas Robert Clifford was able to benefit from a well-established office, which included a small number of royal gunners in receipt of annual wages.[80] The numbers of men employed in ordnance companies also grew massively, with as many as 1,000 men in service in 1497; a noticeable increase on the 605 men recruited for the 1475 expedition.[81] Nevertheless, it is also clear that some things remained the same throughout the period, including the use of commissions to recruit workers for the ordnance and the role of Calais as an arsenal.

79 David Grummitt, 'The Defence of Calais and the Development of Gunpowder Weaponry in England in the Late Fifteenth Century', *War in History*, 7 (2000), 253–72 (p. 270).

80 Note that permanent masters of the ordnance were employed in France between 1432 and 1450 but a similar official did not exist in England until the latter date: see C. T. Allmand, 'L'artillerie de L'armée Anglaise et Son Organisation a L'époque de Jeanne d'Arc', in *Jeanne d'Arc, une Époque, un Rayonnement*, ed. R. Pernaud (Paris: CNRS, 1982), pp. 73–83.

81 Paul Leroy Homer, 'Studies in the Military Organization of the Yorkist Kings' (unpublished doctoral thesis, University of Minnesota, 1977), pp. 71–4.

3

English Royal Ships

In late medieval England, royal ships were the personal property of the king, and their numbers fluctuated markedly over time. A royal navy did not exist as such and vessels were often disposed of to pay debts, as occurred after the death of Edward III in 1377.[1] The construction and maintenance of vessels in the king's fleet were the responsibility of an official known as the clerk or keeper of the king's ships. His duties included the construction, repair, victualling and safe keeping of the royal ships. He was not responsible, however, for the vessels whilst they served in naval operations, which instead were transferred to admirals whose expenses were paid for by treasurers of war. The numbers of royal vessels varied over time, for instance twenty-seven ships were at sea in 1370, twenty in 1372 and thirteen in 1374. These ships only comprised a small proportion of the vessels used in naval expeditions, as the vast majority of ships used were hired or impressed merchant ships. Nevertheless, royal ships served an important role as warships, troop transports and as ships for diplomatic missions.[2] The armament of these vessels changed considerably over the course of the fifteenth century. Henry V's great ship, the *Grace Dieu* of 1,400 tons, completed in 1420, was the largest royal vessel constructed before the seventeenth century, yet it only possessed three guns.[3] By contrast, Henry VII's slightly smaller ship, the *Regent* of 1,000 tons, was equipped with a total of 225 guns for an expedition to Scotland in 1497.[4] The

1 James Sherborne, 'The Hundred Years' War. The English Navy: Shipping and Manpower 1369–1489', *Past and Present*, 37 (1967), 163–75.

2 Ibid., pp. 166–7; Susan Rose, ed., *The Navy of the Lancastrian Kings: Accounts and Inventories of William Soper, Keeper of the King's Ships, 1422–1427* (London: Allen & Unwin for the Navy Record Society, 1982), p. 30; Graham Cushway, *Edward III and the War at Sea: The English Navy, 1327–1377* (Woodbridge: Boydell Press, 2011).

3 W. J. Carpenter-Turner, 'The Building of the Gracedieu, Valentine and Falconer at Southampton, 1416–1420', *The Mariner's Mirror*, 40 (1954), 55–72 (p. 55); Rose, *The Navy of the Lancastrian Kings*, p. 38; TNA, E 364/57, rot. I dorse.

4 M. Oppenheim, ed., *Naval Accounts and Inventories of the Reign of Henry VII, 1485–8 and 1495–7*, Publications of the Navy Records Society (London: Naval Records Society, 1896), p. 274.

intervening seventy-seven years had therefore witnessed a significant change in the use of artillery in naval warfare. This chapter traces how and why this development occurred.

The principal source for the study of royal ships is the accounts of the keeper of the king's ships, which survive regularly for the years 1327–1452 and after a gap of thirty-two years survive intermittently from 1485 onwards.[5] For the earlier period these exist in enrolled form in the foreign account rolls of the Exchequer.[6] The accounts are in a standard format consisting of: the receipt of money, receipt of ships and equipment, purchases, expenditure of equipment and, finally, the remaining ships and equipment in the possession of the keeper. They provide useful information on the number of guns on royal ships but are not without their limitations. There is little evidence for how these weapons were used and in many cases details of gunpowder, gunstones and other equipment are omitted.

After 1452 three further sets of accounts survive for the keeper of the king's ships for the fifteenth century.[7] These are the accounts of Thomas Rogers for the years 1485–8, submitted by his widow Joan, and the accounts of Robert Brigandine for the years 1495–7 and 1498–1500, which take the form of particulars of account. For the years 1453 to 1484, evidence from other Exchequer sources, such as the issue rolls, tellers' rolls and warrants for issue, can be used to bridge the gap in the keeper accounts. Furthermore, the household accounts of John, Lord Howard, contain numerous entries relating to the naval expeditions of 1468 and 1481.[8]

EDWARD III TO RICHARD II

The earliest known evidence for the use of firearms on an English ship occurs in the early fourteenth century. This was for the royal ship, the *All Hallow's Cog*, whose accounts cover the period from March 1337 to September 1338. The expenses section includes 3s. spent upon the purchase of an iron instrument for firing quarrels and lead shot with gunpowder for the defence of the ship.[9] Later, in 1372, twenty-nine iron guns along with lead shot and saltpetre were provided

5 See Cushway, *Edward III and the War at Sea*; Rose, *The Navy of the Lancastrian Kings*, pp. 28–55; Oppenheim, *Naval Accounts and Inventories of the Reign of Henry VII*, p. xv.

6 Particulars of account rarely survive see Rose, *The Navy of the Lancastrian Kings*, p. 3.

7 Oppenheim, *Naval Accounts and Inventories of the Reign of Henry VII*, pp. 1–81, 133–342; TNA, E 36/8.

8 Anne Crawford, ed., *The Household Books of John Howard, Duke of Norfolk, 1462–1471, 1481–1483* (Stroud: Sutton for Richard III & Yorkist History Trust, 1992).

9 Ian Friel, *The Good Ship: Ships, Shipbuilding and Technology in England, 1200–1520* (London: British Museum Press, 1995), p. 152; TNA, E 101/20/27, m. 1d.

by the privy wardrobe for a naval expedition.[10] These appear to have been isolated examples, however, as no further evidence survives for their maritime use prior to the late fourteenth century. It does appear, though, that guns were being used on merchant ships which were impressed for naval expeditions, as can be seen by the presence of a gunner in a fleet commanded by Sir Richard Arundel in 1378.[11]

There is very limited evidence for the provision of guns in the king's keeper of the ships accounts for the reign of Richard II. This consists solely of a single entry for the *Saint Mary of Mortrigo* which was equipped with one gun for a voyage of the admiral Edward Courtenay, earl of Devon in 1384.[12] Three years later, a total of seven gunners were mustered for the naval expedition led by Richard, earl of Arundel.[13] Later in 1398, an inquisition was set up to investigate a complaint made by a Walter Lange that nine guns intended for the defence of one of his ships had been removed from the vessel without his consent.[14] Therefore the limited evidence for the period suggests that guns on ships were growing in importance by the end of the fourteenth century. Yet this did not seem to apply to royal ships, which appear to have been very rarely armed with gunpowder weapons.

HENRY IV

In the reign of Henry IV guns began to assume a more important role in the armament of royal ships. This process can be traced through the accounts of the keepers of the king's ships which survive in an almost unbroken sequence from 1399 to 1452.[15] Upon his accession to the throne in 1399, Henry IV inherited a small number of royal ships, none of which were armed with guns.[16] The accounts for John Chamberlain for the years 1401 to 1405, however, include a list of weaponry supplied to four royal ships: *The Katherine of the Tower*, *The George of the Tower*,

10 T. F. Tout, 'Firearms in England in the Fourteenth Century', *The English Historical Review*, 26 (1911), 666–702 (pp. 675, 693–4); Jonathan Sumption, *The Hundred Years War: Volume III, Divided Houses* (London: Faber, 2009), p. 144.

11 TNA, E 101/36/32, m. 2, from the AHRC-funded database, *The Soldier in Later Medieval England*, http://www.icmacentre.ac.uk/soldier/database/ [Accessed 25 August 2014].

12 TNA, E 364/21, rot. B; E 364/23, rot. B.

13 TNA, E 101/40/33, m. 2, from the AHRC-funded database, *The Soldier in Later Medieval England*, http://www.icmacentre.ac.uk/soldier/database/ [Accessed 25 August 2014].

14 H. C. M Lyte, ed., *Calendar of Inquisitions Miscellaneous*, volume 6 (London: H. M. S. O., 1916–2003), p. 93.

15 Rose, *The Navy of the Lancastrian Kings*, pp. 29–56; the exception is for Elmying Leget who was keeper from June 1411 until March 1413.

16 TNA, E 364/39, rot. F.

The Nicholas of the Tower and *The Gracedieu of the Tower*.[17] All of these arms were delivered by John Norbury, keeper of the privy wardrobe at the Tower of London, to John Chamberlain by indenture. A total of thirteen guns, including one 'double gun', were supplied to these ships along with other equipment, although the accounts do not specify how the weapons were allocated.[18] It appears that these weapons were unused as none of the gunpowder, tampons or ammunition was expended during the time of the accounts.[19] There is also evidence that guns were issued to non-royal ships in this period. The accounts of John Norbury include a list of equipment which is described as having been expended by ships at different times.[20] This consisted of 285lbs of lead for lead shot, 6 guns, 212 gunstones, 300 tampons and 1,238lbs of gunpowder. These items are not included in the accounts for John Chamberlain, which indicates that they were issued directly to the masters of non-royal ships during naval expeditions.

During the time of Chamberlain's successor to the office, John Elmeton, the number of royal ships contracted to only two.[21] When John Starling replaced Elmeton in 1409, however, the number of vessels increased to seven, of which four were equipped with guns.[22] This consisted of two guns on the *Bernard*, two guns on the *Mary of the Tower*, one gun on the *Carrack*, and three guns, one handgun and a barrel of gunpowder on the *Christopher*.[23] The accounts for Starling's successor, Elmying Leget, do not survive, but a comparison with those of William Catton, the first keeper of the king's ships for Henry V, indicate that these four ships were later disposed of.[24] Therefore by the end of the reign of Henry IV, a significant proportion of royal ships were routinely armed with guns. These vessels were used for a variety of purposes, such as transporting Henry IV's second wife, Joan of Navarre, to England in 1403, and in the naval expedition of Thomas Beaufort the following year which patrolled the Straits of Dover.[25] Yet the number of royal ships

17 TNA, E 364/43, rot. E dorse.

18 The meaning of 'double gun' is unclear, but it might refer to a double-barrelled or a double-chambered gun.

19 For definitions, see H. L. Blackmore, *The Armouries of the Tower of London* (London: H. M. S. O., 1976), pp. 242–6.

20 TNA, E 101/404/25.

21 For the period 1405–9, Rose, *The Navy of the Lancastrian Kings*, p. 33.

22 John Starling's accounts run from 1409–11, TNA, E 364/46, rot. E dorse; Alan Moore, 'Accounts and Inventories of John Starlyng, Clerk of the King's Ships to Henry IV', *Mariner's Mirror*, 4 (1914), 20–6.

23 Ibid.

24 William Catton was keeper of the king's ships from 1413–20. Rose, *The Navy of the Lancastrian Kings*, pp. 34, 38.

25 TNA, E 364/43, rots. E–E dorse; G. L. Harriss, 'Beaufort, Thomas, Duke of Exeter (1377?–1426)', in *Oxford Dictionary of National Biography*, http://www.oxforddnb.

fluctuated over the course of the reign and it is unclear if the presence of guns made a notable difference to their fighting capabilities.

HENRY V

The accession of Henry V to the throne in 1413 led to a significant rise in the number of royal ships.[26] William Catton, appointed in July of the same year, took receipt of eight vessels, of which three were equipped with guns.[27] These consisted of the *Thomas of the Tower* with four guns, the *Grand Mary of the Tower* with three guns, and the *Katherine of the Tower* with one damaged gun. The number of royal vessels had doubled to sixteen by the end of Catton's first period of accounts in 1416. There was only a very slight increase in the provision of guns, however, with one further damaged gun being provided for the *Katherine of the Tower* and another damaged gun for the *Red Cog of the Tower*. In addition, two damaged guns remained in the storehouse at Greenwich, together with sixty gunstones received from Simon Fleet, keeper of the privy wardrobe. The period of Catton's second set of accounts, 1416 to 1420, saw the royal fleet reach its apex at thirty-six vessels.[28] Catton was replaced by William Soper, in 1420. At this time, the king owned thirty ships, of which fourteen were armed with guns (see Table 6).[29] This group consisted of the *Trinity Royal* with five guns, the *Holy Ghost of the Tower* with six guns, the *George* with three guns, the *Thomas of the Tower* with four guns, the *Katherine of the Tower* with two guns, the *Nicholas* with one damaged gun, the *Grand Mary of the Tower* with three guns, the *Paul of the Tower* with three damaged guns, the *Christopher of the Tower* with two guns, the *Andrew of the Tower* with two guns, the *Peter of the Tower* with three guns, the *Holy Ghost of Spain* with two guns, the *Grace Dieu* with three guns and the *Roos* with one gun.

Many of these vessels were captured prizes, such as the *George*, the *Paul of the Tower*, the *Christopher of the Tower*, the *Andrew of the Tower* and the *Peter of the Tower*.[30] In addition there were three large ships which were constructed at Southampton by Robert Berde under the supervision of William Soper: the *Holy Ghost of the Tower*, the *Trinity Royal* and the *Grace Dieu*. By the end of Soper's first set of accounts, in 1422, the number of royal vessels had decreased slightly to

---

com/view/article/1864?&docPos=30&backToResults=/search/refine/?docStart=1 [Accessed 25 August 2014].

26 Rose, *The Navy of the Lancastrian Kings*, p. 34.

27 TNA, E 364/54, rot. G.

28 TNA, E 364/57, rot. I; E 364/61, rots. L dorse, M; Rose, *The Navy of the Lancastrian Kings*, p. 37.

29 TNA, E 364/61, rots. I–J.

30 For a definition of a carrack see Rose, *The Navy of the Lancastrian Kings*, p. 255.

twenty-four, of which twelve were equipped with guns.[31] This comprised of the newly constructed ship the *Jesus of the Tower* with two damaged guns, the *Trinity Royal* with five guns, the *Holy Ghost of the Tower* with six guns, the *George* with three guns, the *Thomas of the Tower* with four guns, the *Grand Mary of the Tower* with three guns, the *Paul of the Tower* with one damaged gun, the *Christopher of the Tower* with two guns, the *Peter of the Tower* with three damaged guns, the *Holy Ghost of Spain* with one gun, the *Grace Dieu* with three guns and the *Roos* with one gun. This meant that by the death of Henry V, in 1422, a high proportion of royal ships were armed with guns (54 per cent). There was also a small increase in the number of these weapons used on individual vessels (see Table 6). This increase mainly occurred on the larger ships such as the *Holy Ghost of the Tower* of 760 tons with six guns and the *Trinity Royal* of 540 tons with five guns; although the much smaller *Thomas of the Tower* of 180 tons also had six guns.

The unparalleled expansion in the number of well-armed royal ships was necessary for the naval strategy of Henry V. This included regular patrols of the English Channel by small fleets of ships, particularly in the years 1417 to 1421. These squadrons were able to defeat the French and their allies at sea and helped to facilitate the conquest of Normandy.[32] The accounts of Catton and Soper demonstrate that guns were regularly used in naval engagements. For example we know that 47lbs of gunpowder was purchased and used by the *Holy Ghost of the Tower* at the battle of the Seine in 1416.[33] In the following year, the same ship suffered the loss of one gun, which was captured in a naval battle.[34] On 3 February 1420, the *Thomas of the Tower* expended ten gunstones; later the same year, the *Grace Dieu* fired 100lbs of gunpowder and lost one of its four guns in the defence of the vessel.[35] Yet the accounts generally fail to record information on the gunstones, gunpowder and tampons provided to the individual ships. An exception to this rule can be seen with the *George*, which was provided with three guns, seventy-two gunstones, seventy-two tampons and 65lbs of gunpowder for a naval expedition commanded by John Holland, earl of Huntingdon, in 1417.[36] It is unclear exactly why certain vessels were allocated guns as opposed to others. For instance the carrack the *Mary of Sandwich* of 550 tons was unarmed, whereas the small balinger

31 TNA, E 364/61, rots. I–J.

32 C. F. Richmond, 'English Naval Power in the Fifteenth Century', *History*, 52 (1967), 1–15 (p. 1).

33 Rose, *The Navy of the Lancastrian Kings*, pp. 218–19.

34 TNA, E 364/59, rot. J; Friel, *The Good Ship*, p. 151.

35 TNA, E 364/59, rot. J; E 364/57, rots. I–I dorse; Rose, *The Navy of the Lancastrian Kings*, pp. 50–1.

36 This information is contained in the accounts of William Soper for 1420–2, yet they seem to relate to the naval expedition of 1417, as the earl of Huntingdon was not engaged in naval expeditions in the former years. See TNA, E 364/61, rot. K dorse.

TABLE 6 The Lancastrian navy: the number of guns on specific ships by year.

Type	Tonnage	1409	1411	1413	1416	1420	1422	1427	1432	1436	1439	1442	1447	1452
Ship		2	2											
Ship		2	2											
Ship		3	3											
Ship		1	1											
Ship	180			4	4	4	6							
Ship	126			3	3	3	3							
Ship	210			1	2	2								
Ship	120				1									
Ship	540					5	5	5	5	5	5	5	5	1
Ship	760					6	6	6	6	6	6			
Carrack	500					3	3							
Ship	30					1								
Carrack	600					3	1							
Carrack	600					2	2							

Type	Tonnage	1409	1411	1413	1416	1420	1422	1427	1432	1436	1439	1442	1447	1452
Carrack	400					2	2							
Carrack	600					3	3							
Ship	290					2	1							
Ship	1400					3	3	3	3	3	3			
Balinger	30					2	1							
Ship	540						2	2	3	3	3	3		

Sources: TNA, E 364/43, rots. E–E dorse; E 364/46, rot. E dorse; E 364/54, rots. G–H; E 364/57, rot. I dorse; E 364/59, rots. I–I dorse; E 364/61, rots. I–M; E 364/65, rots. F dorse–G dorse; E 364/69, rots. S dorse, T; E 364/71, rots. D–D dorse; E 364/73, rots. O–O dorse; E 364/76, rots. C–C dorse; E 364/81, rots. G–G dorse; E 364/86, rots. G–G dorse.

the *Roos* of thirty tons had one gun (see Table 6).[37] There does seem to have been a trend though, for the heavier vessels, such as the ships and carracks, to have been provided with guns, as opposed to the smaller balingers. In 1422, 87.5 per cent of the eight ships and 71.4 per cent of the seven carracks had guns, as opposed to 11 per cent of the nine balingers.[38] The ships, on the whole, were supplied with guns provided by the keeper of the king's ships or from the royal arsenal at the Tower of London, whereas the weaponry for the carracks was obtained when they were captured in 1416 and 1417.

HENRY VI

The death of Henry V led to the rapid disposal of most of the king's vessels as his will stated that the royal ships should be sold to pay off some of his debts. These sales also had the advantage of reducing expenditure at a time when control of the English Channel looked assured, because Normandy was firmly under English control.[39] By 1427 only two balingers and the four largest ships remained. The latter were armed with guns and comprised the *Grace Dieu* with three guns, the *Holy Ghost of the Tower* with six guns, the *Jesus of the Tower* with two damaged guns and the *Trinity Royal* with five guns.[40] These ships did not see active service again; instead they were docked on the mud flats in the Hamble.[41] In addition, no attempt was made to increase the weaponry of these ships with guns left over from the sales of previous vessels. Instead the guns were gradually given away or sold off. For instance six guns were given to the treasurer of Calais on 27 January 1439 and nine more guns were given to three esquires of the king in July 1449.[42] Soper was replaced by Richard Clyvedon in 1442, who continued to dispose of old stores left in the office. By 1452 none of the royal ships remained, with the *Grace Dieu* having been destroyed by lightning in 1439 and the other vessels scrapped.[43] The crown possessed very few royal vessels after 1452 and there is little extant information on how they were equipped. One of the few relevant documents we possess is a commission given to a certain John Hadilsey, on 29 March 1460, to seize various

37 For a definition of a balinger see Rose, *The Navy of the Lancastrian Kings*, p. 253; C. F. Richmond, 'The Keeping of the Seas during the Hundred Years War: 1422–1440', *History*, 49 (1964), 283–98 (p. 287).

38 TNA, E 364/61, rots. I–J.

39 Rose, *The Navy of the Lancastrian Kings*, p. 52; Ralph A. Griffiths, *The Reign of King Henry VI: The Exercise of Royal Authority, 1422–1461* (London: Benn, 1981), p. 179.

40 TNA, E 364/69, rot. S dorse.

41 Rose, *The Navy of the Lancastrian Kings*, pp. 53–4.

42 TNA, E 364/76, rot. C; E 364/86, rots. G–G dorse.

43 Rose, *The Navy of the Lancastrian Kings*, p. 55.

habiliments of war including guns and gunpowder for the defence of a ship called the *Grace Dieu*.[44]

From the 1430s onwards, the English coastline and merchant vessels came under renewed threat of attack, following French victories in Normandy and the defection of Philip the Good, duke of Burgundy, in 1435.[45] But this did not prompt a new programme of royal shipbuilding. Instead the government of Henry VI relied almost entirely upon merchant shipping to keep the sea.[46] There is some evidence of guns assuming a more important role in warfare at sea during these years. For example, it was reported that the *Isabella of Calais*, whilst on a voyage to Calais, was broken and battered as a result of gunfire from enemy ships and driven onto the rocks outside Dover Castle in 1426.[47] Later in 1441, a commission of oyer and terminer was appointed to look into claims that a ship called the *Saint Nicholas* was attacked with weapons including gunpowder weapons.[48] In 1449, Robert Winnington reported that during a naval battle with a Hanseatic fleet of 100 ships, the English were attacked with 1,000 guns.[49] The latter example also suggests that there may have been an increase in the number of guns used on ships. This is corroborated by a petition made by a Thomas Martin, in which he mentioned that his ship, the *Mary Buckingham*, was strengthened with nine guns and four barrels of gunpowder, after being arrested for the expedition of John Talbot, earl of Shrewsbury, to Gascony in 1453.[50] Further evidence for the frequency with which guns were used in naval combat in the mid-fifteenth century can be seen from a petition dating to 4 December 1460. This document states that William Staveley, who had served in the Calais garrison for fifteen years and had 'been often tymes sore wounded and hurt' in the king's service 'upon the see', had been injured by a gun whilst taking part in the attack made by the captain of Calais, Richard Neville, earl of Warwick, on Spanish ships in the summer of 1459.[51] There is also limited evidence that the government on occasion provided ordnance for naval expeditions. For example, on 26 June 1442 warrants were given for the delivery of eight barrels of gunpowder to a fleet appointed to keep the sea.[52] Later in August 1458, 100 marks were given to Lord Fauconberg for the purchase of artillery for

44 This was a different ship from that of Henry V. Richmond, 'English Naval Power in the Fifteenth Century', p. 7; *CPR 1452–1461*, pp. 554–5.

45 Griffiths, *The Reign of King Henry VI*, pp. 423–4.

46 Richmond, 'The Keeping of the Seas during the Hundred Years War', pp. 291–3.

47 *CPR 1422–1429*, p. 385.

48 *CPR 1436–1441*, p. 575.

49 Griffiths, *The Reign of King Henry VI*, p. 431.

50 *CPR 1452–1461*, p. 76.

51 TNA, C 81/1477, no. 31.

52 Harris Nicolas, ed., *Proceedings and Ordinances of the Privy Council of England*, volume 5 (London: Record Commission, 1834–7), p. 190.

ships.[53] Therefore, despite the disposal of the royal fleet of vessels and the limited provision of ordnance provided to maritime expeditions, guns continued to grow in importance to naval warfare during the reign of Henry VI.

EDWARD IV TO RICHARD III

Under Edward IV the numbers and types of guns used on ships changed significantly. Coinciding with these developments was the gradual reconstruction of a royal fleet of vessels. In his early years, Edward IV was reliant upon the impressment of merchant ships or the support of magnates with large fleets, such as the earl of Warwick and Lord Howard. Upon Edward IV's accession to the throne in 1461, he inherited the *Grace Dieu* and soon afterwards purchased the large ship the *Margaret of Orwell* from some merchants of Ipswich.[54] It appears that Edward appreciated the importance of naval power, as efforts were soon made to furnish these ships with large quantities of ordnance. On 20 June 1461, a warrant was issued for the payment of William Baldry, clerk of the *Margaret of Orwell*, who had purveyed 1,000lbs of gunpowder and 600 gunstones for the furnishing of that ship, with further quantities of guns, gunstones and gunpowder ordered four months later.[55] In the same year, another warrant was issued for the delivery of twelve barrels of gunpowder to the ships of the king.[56] The earliest recorded evidence for a gunner on English ships in the fifteenth century occurs in 1462. In July of that year, John Smith, gunner, was paid for having served in two ships of the king, the *Bage* and the *Margaret of Orwell*.[57] The number of guns on these royal ships is unknown, although a ship of Bayonne, which was delivered by William Eustace, sergeant of arms, to Arnold Trussell, was described as having eleven guns, great and small, in 1466.[58]

Later in 1468, the naval expedition to Brittany commanded by Anthony Woodville was supplied with 2,900lbs of gunpowder.[59] Lord Howard was tasked with assembling the east coast contingent of the fleet and his household accounts reveal that five gunners of Southwark were employed on these ships.[60] Limited

53 TNA, E 404/71/2, no. 79.

54 This was also referred to as the *Margaret of Ipswich*. Richmond, 'English Naval Power in the Fifteenth Century', p. 13; it was referred to as 'great' in TNA, E 404/72/1, no. 17.

55 TNA, E 404/72/1, nos. 17, 53; E 403/824, m. 7.

56 The warrant for issue has the date of 27 July crossed out, so it is unclear when it was issued, TNA, E 404/72/1, no. 12.

57 TNA, E 403/825, mm. 8, 10.

58 TNA, E 101/55/3.

59 TNA, PSO 1/32, no. 1682.

60 Anne Crawford, *Yorkist Lord: John Howard, Duke of Norfolk, c.1425–1485* (London: Continuum, 2010), p. 54; Crawford, *The Household Books of John Howard*, I, 574–5.

preparations were made to counter the threat posed by the earl of Warwick in June 1470, with two barrels of gunpowder being purchased for the keeping of the sea.[61] Greater efforts were made to furnish the fleet tasked with recapturing St Michael's Mount in Cornwall from John de Vere, earl of Oxford, in 1473.[62] Indentures made with the captains of four royal ships, the *Caricon, Christopher of Calais, Mary of Calais* and *Garce*, specified that they should be supplied with gunpowder and weaponry at the king's expense.[63] In the same year, £41 13s. 4d. was paid to Egidio van Rasynghin, master gunner of the king, in full payment of thirty-nine guns for the furnishing of the *Caricon*.[64] This meant that Edward IV's ship the *Caricon* had more guns in 1473 than Henry V's entire fleet of royal vessels in 1422 (see Table 8). Other ships were also equipped with sizeable numbers of guns by the 1470s. In 1479, Lord Howard reported that two of his ships had been captured, the *George Howard* with sixteen guns and the *Edward Howard* with fifteen guns.[65]

War with Scotland in 1480 prompted the appointment of Thomas Rogers as clerk of the king's ships.[66] In the same year, one barrel of gunpowder was given to the royal ship the *Antony*.[67] In 1481, Lord Howard was appointed as commander of a fleet of ten ships tasked with attacking the coast of Scotland, with a further fleet of six ships assigned to 'kepe the narowe see'.[68] This led to substantial expenditure by the Exchequer for equipping the king's fleets. Thus Thomas Rogers paid £13 16s. 8d. for the purchase of twelve new guns and for the repair of two damaged guns, with further unspecified quantities of handguns, ship guns and other guns being ordered from Flanders at the price of £66 13s. 4d.[69] The *Mary of Greenwich* was provided with one barrel of gunpowder, 400 gunstones, tampons and other items at a total cost of £22 17s.[70] The *Caricon* was also supplied with 354lbs of gunpowder

61 Crawford, *The Household Books of John Howard*, p. xxiv; TNA, E 364/110, rot. A dorse.

62 Cora L. Scofield, *The Life and Reign of Edward the Fourth: King of England and of France and Lord of Ireland*, volume 2 (London: Longmans, Green, 1923), pp. 59–61.

63 TNA, E 101/71/5, nos. 15–16.

64 TNA, E 405/57, m. 7d.

65 Ian Friel, 'Oars, Sails and Guns: The English and War at Sea, c.1200–c.1500', in *War at Sea in the Middle Ages and Renaissance*, ed. John B. Hattendorf and Richard W. Unger (Woodbridge: Boydell Press, 2003), p. 73; TNA, C 76/163.

66 A position which he was to hold until his death in 1488. Richmond, 'English Naval Power in the Fifteenth Century', p. 10.

67 TNA, E 405/68, m. 3r.

68 Richmond, 'English Naval Power in the Fifteenth Century', p. 10; Crawford, *The Household Books of John Howard*, p. xxiv, II, 2.

69 TNA, E 405/566, ff. 62–3.

70 Otherwise known as the *Great Spaniard*. Crawford, *The Household Books of John Howard*, II, 1; TNA, E 405/566, f. 64.

in two barrels at the price of £11 7s. 6d.[71] A further £73 14s. was paid for 2,752lbs of gunpowder, with £40 more spent on guns, gunpowder and other items for ships.[72]

The household accounts of Lord Howard also reveal that his ships were well equipped with ordnance for the expedition. A total of seventy-one serpentines were purchased from Cornelis, a gunmaker of Saint Katharine in London, Nicholas, a smith of Wivenhoe in Essex, and an unnamed armourer of 'Paul's Cheyne' for £12 13s. 4d.[73] Ten of these serpentines were each described as weighing 114lbs together with their three chambers.[74] Nicholas also forged guns weighing 422lbs for the newly constructed ship the *Barbara*.[75] Further amounts were expended upon the stocking of these guns and in providing them with gunpowder.[76] Five guns of the ship the *Jenet*, together with one other gun and two little broken guns, were also noted as being stored in a churchyard at Harwich.[77] In addition to this, five gunners were paid wages by Lord Howard. Two of them served on a ship of Edward Brampton called the *Great Carvel of Portugal*, with one other gunner employed on the *Trinity of Saint Osyths*.[78] Weaponry was also supplied to the fleet by the garrison of Calais. An indenture made between William Comersale, master of the *Michael of the Tower*, and William Rosse, victualler of Calais, on 12 June, specified that the latter would transfer equipment including 7 large guns, 100 hackbuts, 150 handguns and 5 lasts of gunpowder.[79] It is not possible to determine exactly how many guns were deployed on the king's ships in 1481. Yet the evidence from the Exchequer and the household accounts books of Lord Howard suggests that they were well armed with a variety of guns, including serpentines and handguns.

Naval activity continued into the reigns of Edward IV's two immediate successors, his son Edward V, and his brother Richard III. In 1483, Thomas Rogers was responsible for provisioning the fleet of Sir Edward Woodville, which included expenditure of £304 15s. on gunpowder and saltpetre, together with £226 7s. 7d.

71 TNA, E 405/566, f. 76.

72 Ibid., ff. 29, 49.

73 For a definition of serpentines see Blackmore, p. 242; Crawford, *The Household Books of John Howard*, II, 19, 33, 35, 47, 50.

74 Ibid., II, 47.

75 Ibid., II, 67.

76 Ibid., II, 38, 43, 47.

77 This was a ship confiscated for piracy in 1476, Crawford, *Yorkist Lord*, p. 84; Crawford, *The Household Books of John Howard*, II, 23.

78 Ibid., II, 249, 263, 269, 271–2.

79 Also referred to as the *Nicholas of the Tower*, Richmond, 'English Naval Power in the Fifteenth Century', p. 14. For definitions of hackbuts and handguns see Appendix J. David Grummitt, 'The Defence of Calais and the Development of Gunpowder Weaponry in England in the Late Fifteenth Century', *War in History*, 7 (2000), 253–72 (p. 266); Scofield, *The Life and Reign of Edward the Fourth*, volume 2, pp. 314–15; TNA, E 101/198/13, f. 6r.

on ordnance 'that is bought and most nedis over the kynges own stuffes delyverd out of the Towre'.[80] The following year, he purchased twenty new guns and two serpentines at Southampton for the king's ships.[81] By the time of the accession of Henry VII to the throne in 1485, the royal fleet of vessels consisted of four armed ships. They comprised the *Grace Dieu* with twenty-one damaged guns, the *Mary of the Tower* with forty-eight guns and twelve hackbuts, the *Martin Garsia* with thirty guns and four handguns, and the *Governor* with seventy guns and seven hackbuts.[82] The use of guns on ships was therefore transformed in the reign of Edward IV, particularly during his second reign from 1471 onwards.

HENRY VII

The number of vessels in the royal fleet declined slightly during the reign of Henry VII, from seven in 1485 to five in 1488, a number at which it remained until Henry's death in 1509.[83] The *Martin Garsia* was gifted to Sir Richard Guildford, master of the king's ordnance, on 24 December 1485, and the *Governor* was transferred to two merchants of London on 18 February 1486. The *Grace Dieu* appears to have been in poor condition, as it was listed as having twenty-one damaged guns in 1485, ten of which were transferred to the *Mary of the Tower*, in addition to one barrel of gunpowder delivered to Matthew Baker, esquire for the king's body, on 25 February 1486. The *Grace Dieu* was subsequently transferred with its remaining equipment to Sir Reynold Bray to be broken up for the construction of the *Sovereign*, for which work had begun in 1486.[84] Of the vessels Henry VII inherited in 1485, only the *Mary of the Tower* was retained as an armed royal ship by the end of Thomas Roger's accounts in 1488. The ship was hired by Sir William Capell and other merchants of London for a voyage to Lombardy on 28 February 1486, with 704lbs of gunpowder provided by Roger, which was purchased from Southampton and London. The *Mary of the Tower* was returned to the clerk's control on 30 July 1487, with three guns, five gun chambers, four barrels of gunpowder and one gun hammer

80 Rosemary Horrox, ed., 'Financial Memorandum of the Reign of Edward V: Longleat Miscellaneous Manuscript Book II', *Camden Miscellany*, XXIX. Camden 4th ser. XXXIV (London: Offices of the Royal Historical Society, 1987), pp. 219–20.

81 Rosemary Horrox and P. W. Hammond, eds., *British Library Harleian Manuscript 433. Vol. 2, Second Register of Richard III* (Gloucester: A. Sutton for the Richard III Society, 1980), p. 112.

82 Note that the *Caricon* had been renamed the *Mary of the Tower* in 1479. Oppenheim, *Naval Accounts and Inventories of the Reign of Henry VII*, pp. 36, 50, 69, 73.

83 Richmond, 'English Naval Power in the Fifteenth Century', pp. 11–12.

84 Oppenheim, *Naval Accounts and Inventories of the Reign of Henry VII*, pp. 34, 38, 42, 47, 70, 74.

having been expended, broken or lost during the voyage. The remaining 200lbs of gunpowder and fifty-five guns were then transferred to the king' storehouse at Greenwich, with only the twelve hackbuts left on board. The storehouse itself had twenty-six damaged guns on 4 October 1485.[85]

Ordnance was subsequently transferred from there on 14 February 1488 to arm merchant vessels in the fleet of Sir Charles Somerset, Admiral of the King's Army upon the Sea. This included six guns delivered to the *John of Plasgencia* of 220 tons, nine guns to the *Peter Deneta* of 240 tons and a further nine guns to the *John de Hoyo* of 140 tons. Six days later two men were paid for their expenses in purchasing equipment and for the arresting of guns for the king from various ships in the Thames. This measure could have been taken to arm other merchant ships in the king's service, such as the five ships retained from 9 April 1487 until 10 May 1488.[86] It is clear from surviving accounts, however, that private vessels were equipped with guns for their own protection when sent on trading missions. In 1485, purchases for the provisioning of the *Margaret Cely* for a voyage to Zealand included four small serpentines, two large serpentines, two handguns, two other guns, lead shot, gunstones and gunpowder.[87] On occasion merchant ships also hired gunners, with the *Trinity of Bristol* and *George of Bristol* each employing one gunner for a voyage to North Africa in 1480–1.[88] There is no evidence in the wages section of the accounts of Roger, however, for the employment of gunners on any ships, including during wartime in 1487 and 1488.[89] During the first three years of the reign of Henry VII, therefore, the emphasis appears to have been more upon the arming of merchant ships than royal vessels in times of war with France and Scotland. Yet the crown did replace many of the vessels scrapped or given away after 1485 through the construction of new ships. The two largest vessels consisted of the *Sovereign*, finished in 1488, and the *Regent* which was completed in 1490.[90] In early 1488, a total of £257 10s. was paid by Sir Richard Guildford, master of the king's ordnance, for equipping the *Sovereign* with guns.[91] The following year, twenty-four guns were purchased at the price of £34 by Guildford for the *Regent*, ten of which were acquired from the patron of a Genoese carrack, with a further

85 Ibid., pp. 13–14, 55–65, 74.

86 Ibid., pp. 28–30, 32, 78.

87 Henry Elliot Malden, ed., *The Cely Papers: Selections from the Correspondence and Memoranda of the Cely Family, Merchants of the Staple, A.D. 1475–1488* (London: Longmans, Green, and Co., 1900), pp. 177–9, 184.

88 T. F. Reddaway and Alwyn A. Ruddock, eds., 'The Accounts of John Balsall, Purser of the Trinity of Bristol, 1480–1481', *Camden Miscellany*, XXIII. Camden 4th ser. VII (London: Offices of the Royal Historical Society, 1969), pp. 18, 24.

89 Oppenheim, *Naval Accounts and Inventories of the Reign of Henry VII*, pp. 20–30.

90 Richmond, 'English Naval Power in the Fifteenth Century', p. 14.

91 TNA, E 405/76, mm. 4d, 5d.

£16 12s. subsequently being spent on more ordnance for the same ship.[92] Later in 1490, gunpowder, bows and arrows to the value of £418 8s. 4d. were supplied to ships in Portsmouth for the expedition to Brittany, and in the following year, ordnance and gunpowder were supplied to equip a fleet of fourteen ships, which included the *Regent* and the *Sovereign*.[93]

Thomas Rogers was succeeded as clerk by William Combersale in 1488, but the latter's accounts do not survive. The next set of extant documentation comes from Robert Brigandine, who replaced Combersale in 1495.[94] His accounts include the earliest recorded breakdown of the location of guns on an English ship. This survey was drawn up for the *Sovereign* whose inventory listed it as possessing 141 guns, consisting of 110 serpentines and 31 stone guns.[95] Of these, 16 serpentines were located in the forecastle above the deck, 24 serpentines in the forecastle below, 20 stone guns in the waist of the ship, 4 serpentines in the stern, 25 serpentines in the deck over the somercastle, 20 serpentines of iron in the poop deck, together with 21 serpentines and 11 stone guns in the somercastle (see Figure 1).[96] These weapons were also supplied with miches and forelocks, together with 5 barrels of gunpowder, 400 lead shot, 200 iron dice, 12 gun hammers, 1 ladle for melting lead for lead shot and 6 moulds of stone to cast lead shot.[97] The *Sovereign* was moored at a dock at Erith in the River Thames from October 1495 until March 1496. For the period from 24 October to 13 December, the crew consisted of twenty-eight men, including a Hugh Gonner, whereas from 13 December until 14 March, the crew was reduced to twenty, including a Hans Fryse, gunner. The ship was subsequently conveyed to Portsmouth from 14 March to 15 April 1496, during and after which there is no mention of a gunner in the crew. The *Sovereign* was later hired out to merchants for a voyage to the Levant on 31 January 1497.[98]

The *Regent* was listed as having 181 guns in 1495, but it is not specified where they were located on the ship.[99] Two years later, the weaponry on the ship was increased and refurbished for the impending war with Scotland. Forty-four

92 TNA, E 36/124, ff. 50r, 57r; E 405/77, m. 6r; E 405/78, m. 19d.

93 TNA, E 36/124, f. 52r; E 405/78, m. 20r; WALMR, 12,240, mm. 3, 6, 20, 25–6.

94 Richmond, 'English Naval Power in the Fifteenth Century', p. 12.

95 Oppenheim, *Naval Accounts and Inventories of the Reign of Henry VII*, pp. 194–5, 204–5, 216–17. The name stone gun refers to the type of ammunition used for these weapons, which are likely to have been fowlers (see Appendix J).

96 For a discussion of castles on ships see Friel, *The Good Ship*, pp. 79–81.

97 For a definition of miches see Blackmore, *The Armouries of the Tower of London*, p. 234; whereas for a definition of iron dice see C. S. Knighton and D. M. Loades, eds., *The Anthony Roll of Henry VIII's Navy* (Aldershot: Ashgate Publishing Limited, 2000), p. 13.

98 Oppenheim, *Naval Accounts and Inventories of the Reign of Henry VII*, pp. 161–70, 194–5, 204–5, 216–18.

99 Ibid., p. 261.

FIGURE 1 Location of guns on the *Sovereign* in 1495.

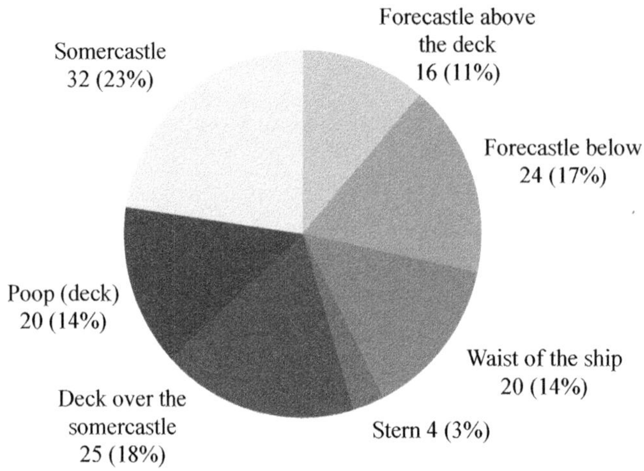

Source: Oppenheim, *Naval Accounts and Inventories of the Reign of Henry VII*, pp. 194–5.

serpentines, together with other equipment, were purchased from the *Mary of the Tower* for the *Regent*.[100] The crayer, the *Trinity of Erith*, was subsequently hired for the conveyance of these guns from the dockyard at Portsmouth to the *Regent*, which was located 'without Portsmouth haven at Stokke Bray', from 28 February until 27 May 1497.[101] Further payments were made to smiths and carpenters for miches, forelocks, bolts and the stocking of guns. At the beginning of the Scottish campaign, therefore, the arsenal of the ship consisted of 195 iron serpentines, 30 brass serpentines, 1 last of gunpowder, 950lbs of lead, 300lbs of iron dice, half a barrel of lead shot and iron dice and 3,000 tampons. The *Regent* appeared to have experienced heavy action during the voyage, as 30 of the serpentines, 8 barrels of gunpowder, 850 weight of lead, 275 iron dice, 2,700 tampons and 25 gunstocks were expended, broken or worn out 'for the use defence and safeguard of the said ship'. Surprisingly, however, the ship does not appear to have employed a gunner

[100] The former appears to have been disposed of by the crown at some point before 1495. Ibid., p. 247.

[101] For a definition of a crayer see Maryanne Kowaleski, 'Warfare, Shipping, and Crown Patronage: The Impact of the Hundred Years War on the Port Towns of Medieval England', in *Money, Markets and Trade in Late Medieval Europe: Essays in Honour of John H.A. Munro*, ed. Lawrin Armstong, Ivana Elbl and Martin M. Elbl (Leiden and Boston: Brill, 2007), pp. 233–56 (p. 249).

during the voyage to Scotland, although a gunner was paid wages from 1 May 1495 to 19 February 1496.[102]

Two other royal ships took part in the 1497 expedition. These consisted of the newly constructed barks, the *Mary Fortune* and *Sweepstake*.[103] The accounts do not state how many guns were deployed on each vessel, although each of them was provided with half a barrel of gunpowder, 50lbs of lead, 46lbs of iron dice and 300 tampons by Sir Robert Clifford, master of the king's ordnance. The *Mary Fortune* subsequently expended or lost these supplies during the course of the voyage. Further quantities of gunpowder, lead shot, iron dice and tampons obtained from the Tower of London were delivered by Robert Brigandine, clerk of the king's ships, to the *Antony of Saltash, Henry of Bristol, Mary Bride of Bristol, Mary Tower of Bristol, Andrew of Plymouth, Michel of Dartmouth* and *Bark of Penzance* (see Table 7).[104] Brigandine only seems to have been responsible for supplying the ships at Portsmouth, however, as additional equipment was directly sent from the Tower of London to ships at the Downs and elsewhere. This consisted of quantities of gunpowder, iron dice and tampons sent to the *Antony of Berkeley, Christopher Lyme, Marlion, Anne Clerk, Mary Rose, John Hampton, God's Grace* and five other unnamed ships. A further 6,400 tampons together with 1 last, 7½ barrels and 1 firkin of gunpowder were also sent to unspecified ships at the Downs.[105] There is little surviving evidence for the ships of Henry VII after 1497. Robert Brigandine's accounts for 1498–1500 are badly damaged, but include payments for repairs to a port piece and serpentines.[106] His accounts for the years 1500–2 are also in poor condition, but contain an inventory which shows that the *Regent* had 181 guns, consisting of 151 iron serpentines and 30 bronze serpentines, together with 1 barrel of gunpowder, 100lbs of lead, 25lbs of iron dice and 300 tampons.[107] The period from 1488 to 1497 therefore saw significant changes in the armament of royal ships, yet the extensive provisioning of non-royal vessels for naval expeditions also remained of great importance.

102 Ibid., pp. 218–23, 247, 250, 279.

103 For a definition of a bark see Rose, *The Navy of the Lancastrian Kings*, p. 254.

104 Oppenheim, *Naval Accounts and Inventories of the Reign of Henry VII*, pp. 303, 312, 329, 334–5, 340–3.

105 TNA, E 36/8, ff. 99v, 106v, 109v.

106 For a definition of a port piece see Blackmore, *The Armouries of the Tower of London*, p. 239; TNA, E 36/6, ff. 30–1.

107 TNA, E 315/317/1, f. 15r.

Table 7 Equipment delivered to ships for the 1497 expedition to Scotland.

Name of ship	Type of ship	Gunpowder	Lead shot (lbs)	Iron dice (lbs)	Tampons
Regent	Royal	1 last	950	300	3,000
Sweepstake	Royal	1/2 barrel	50	46	300
Mary Fortune	Royal	1/2 barrel	50	46	300
Antony of Saltash	Non-royal	1/2 barrel	50	250	250
Henry of Bristol	Non-royal	3.5 barrels	400	200	1,200
Mary Bride of Bristol	Non-royal	2 barrels	200	100	500
Mary Tower of Bristol	Non-royal	2 barrels	200	100	500
Andrew of Plymouth	Non-royal	1.5 barrels	150	750	500
Michel of Dartmouth	Non-royal	2 barrels, 1 firkin	250	125	700
Bark of Penzance	Non-royal	1/2 barrel, 1 firkin	50	250	300
Antony of Berkeley	Non-royal	2 barrels		75	500
Christopher Lyme	Non-royal	1.5 barrels			
Marlion	Non-royal	2 barrels			
Anne Clerk	Non-royal	1 barrel			

Name of ship	Type of ship	Gunpowder	Lead shot (lbs)	Iron dice (lbs)	Tampons
Mary Rose	Non-royal			100	500
John Hampton	Non-royal			150	500
God's Grace	Non-royal			100	
Unspecified ships	Non-royal			345	5,000
Ships in the Downs	Non-royal	1 last, 7.5 barrels, 1 firkin		951	6,400
Total		2 lasts, 28 barrels and 2 firkins	2,350	3,888	20,450

Source: Oppenheim, *Naval Accounts and Inventories of the Reign of Henry VII*, pp. 340–3; TNA, E 36/8, ff. 99v, 106v, 109v.

CONCLUSION

Guns were first used on ships in the fourteenth century but rarely on royal vessels. The reigns of Henry IV and Henry V were significant in that the arming of royal ships with small numbers of guns became widespread. The king's ships were disposed of by the government of Henry VI, but guns remained important in the armament of merchant vessels. The most important development occurred in the reign of Edward IV with the great rise in the number of guns on royal ships, which meant that, for the first time, they were far better armed than merchant vessels. Henry VII continued the policy of having a small number of heavily armed ships, with the period 1488–97 marked by a tremendous increase in the quantity of guns (see Table 8). Royal interest in making use of naval power was crucial in driving these changes, with English kings, such as Henry V and Edward IV, well known for their interest in ordnance.[108] Monarchs were still dependent in large part on the use of merchant ships in wartime, however. At times, such as in 1405 and 1488, private vessels were supplied directly with guns; on other occasions, for instance in 1497, they were only given gunpowder and ammunition.

The period was also characterised by technological changes in the types of guns used in England. Prior to the second half of the fifteenth century, there appears to have been little diversity in the types of guns used on ships. The vast majority of weapons were described interchangeably as guns or cannons, which suggests that they were of the same type without any need to distinguish between them. The accounts of the treasurer of Calais state that six fowlers were transferred by William Soper to the territory in 1437, which suggest that this was the type of gun used on royal ships in the first half of the fifteenth century.[109] Fowlers fired gunstones and were provided with one or more gun chambers. The presence of a large 'double' gun in 1405 and a handgun in 1409 appear to have been isolated examples. The description of a ship of Bayonne as having eleven guns great and small in 1466 suggests that important changes were occurring in the late fifteenth century. By 1481 this included the use of large numbers of serpentines, hackbuts, handguns and 'ship guns'. The use of the term 'ship guns' indicates that some guns were becoming adapted specifically for naval warfare. At the end of the century, however, the *Regent* was solely armed with serpentines, which suggests that this type of gun was considered the most useful in combat. The types of ammunition also changed, with lead shot and iron dice used for serpentines instead of gunstones.

[108] For instance, for Henry V see Christopher Allmand, *Henry V* (New Haven: Yale University Press, 1997), p. 216; for Edward IV and Henry VII see Grummitt, pp. 259, 270.

[109] TNA, E 364/72, rot. M dorse.

The most significant technological development was the invention of the 'miche', which was a swivel mount for guns.[110] The earliest recorded date for miches used on an English ship is for the *Trinity of Bristol* in 1480–1. They are first listed on royal ships in 1495 although there are indications that they were used on royal vessels long before this date. In 1481, ten of the serpentines were described as being 114lbs in weight which would have made them suitable as swivel guns.[111] Furthermore the use of miches onboard ships may provide an explanation for the use of the term 'ship guns' in the Exchequer accounts for the same year. It was also possibly the reason why the *Caricon* was so heavily armed in 1473, with the invention of the miche allowing a larger quantity of guns to be used on ships than had previously been feasible. This appears to be borne out by the deployment of guns in the *Sovereign* in 1495, with miches allowing these weapons to be placed throughout the ship, particularly in the somercastle and forecastles, as they did not have to be stocked or mounted on wooden carriages.[112] Further technological developments were to occur in the sixteenth century, most notably the invention of the gunport, which allowed heavier guns to be deployed on ships.[113] The innovation of the miche, however, was a necessary prerequisite for these changes without which the heavily armed ships of the navy of Henry VIII would not have been possible.

Surprisingly, these technological developments seem to have had a limited effect on personnel in naval warfare. Gunners were employed on ships in the late fourteenth century, but there is no evidence for this in the first half of the fifteenth century. In the late fifteenth century gunners were present on ships, such as during the 1468 and 1481 expeditions, yet in relatively small numbers. By contrast, the navy of Henry VIII in 1546 included a substantial number of gunners organised into teams on the largest ships.[114] This suggests that gunners were still not considered essential for the operation of guns on ships and that the role was still developing by the end of the fifteenth century. Assessing the effectiveness of guns on fifteenth-century English ships is also difficult due to the limited evidence available. It is clear that they were used in naval warfare, as can be seen by the accounts of William Catton, which record the firing of guns by the *Holy Ghost of the Tower*, *Grace Dieu* and *Thomas of the Tower* in 1416, 1417 and 1420. The accounts only provide this type of information incidentally, however, so it is not possible to determine exactly how they were used and how frequently. The small number of guns on ships, including the use of guns in poor condition, and the low proportion of gun

110 Blackmore, *The Armouries of the Tower of London*, p. 234; Friel, 'Oars, Sails and Guns', p. 73.

111 Swivel guns tended to be small in size. Knighton and Loades, *The Anthony Roll of Henry VIII's Navy*, p. 12.

112 See Figure 1.

113 Friel, 'Oars, Ships and Guns', p. 79.

114 Knighton and Loades, *The Anthony Roll of Henry VIII's Navy*, pp. 13, 41–105.

chambers suggest that they had a limited impact on naval warfare in the early part of the fifteenth century (see Table 8). Yet the prominence accorded to the use of guns in petitions relating to piracy indicates that they gradually began to grow in importance throughout the course of the century.

The increase in the number and types of guns on royal ships from 1473 to 1499 demonstrates that these weapons had become essential to naval warfare. In addition to this, the 1480s and 1490s saw a rise in the proportion of gun chambers to guns, undoubtedly with the intention of increasing the potential rate of fire, as can be seen with the *Regent* in 1497 (see Table 8). It has been suggested that one possible reason for the use of guns in large numbers on ships was due to their unreliability, as can be seen by the breaking or loss of twenty-nine of the *Regent's* serpentines that year.[115] Instead this example illustrates the important role of guns in naval warfare by the end of the century. The loss of thirty serpentines represented only 13.3 per cent of the total of 225 guns; furthermore the expenditure of almost all of the ship's store of munitions indicates that the *Regent* saw heavy fighting in 1497. Additional evidence for this can be seen from the extensive provisions provided for the fleet of over seventeen vessels that year, including very large quantities of gunpowder, ammunition and tampons (see Table 8). Traditional weapons, such as bows and spears, continued to be used in large numbers, with guns still intended for targeting enemy personnel as opposed to causing structural damage to other vessels.[116] Yet the great expansion in the numbers of guns on royal ships during the reign of Edward IV was a significant development which had long term implications for naval warfare. It began a process of continued changes in the numbers and types of guns used on royal warships which continued into the sixteenth century and beyond. This meant that the fighting capabilities of the royal warships of the late fifteenth and early sixteenth centuries were greatly altered from that of their predecessors. Warfare at sea was therefore transformed during the fifteenth century.

115 Friel, 'Oars, Ships and Guns', p. 77.

116 For other weapons on the *Regent* in 1497 see Oppenheim, *Naval Accounts and Inventories of the Reign of Henry VII*, p. 274.

TABLE 8 Number of guns on ships by year.

Year	Type of ship	No. of ships	No. of guns	No. of gun chambers	Ratio of chambers to guns	No. of handguns
1337	Royal	1	1			
1384	Royal	1	1			
1398	Non-royal	1	9			
1405	Royal	4	13			
1409	Royal	4	8	8	1.1	2
1413	Royal	3	8	22	2.75	
1416	Royal	4	10	22	2.2	
1420	Royal	14	41	68	1.65	
1422	Royal	13	38	68	1.78	
1427	Royal	4	16	24	1.5	
1453	Non-royal	1	9			
1466	Non-royal	1	11			
1473	Royal	1	39			
1485	Royal	4	169	550	3.25	23
1488	Royal	1	55	135	2.45	
1495	Royal	2	322	812	2.52	
1497	Royal	1	225	582	2.58	
1500	Royal	1	181	582	3.21	

Sources: TNA, E 101/20/27, m. 1d, E 364/21, rot. B; E 364/23, rot. B; E 364/43, rots. E–E dorse; E 364/46, rot. E dorse; E 364/54, rots. G–H; E 364/57, rot. I dorse; E 364/59, rots. I–I dorse; E 364/61, rots. I–M; E 364/65, rots. F dorse–G dorse; E 364/69, rots. S dorse, T; E 364/71, rots. D–D dorse; E 364/73, rots. O–O dorse; E 364/76, rots. C–C dorse; E 364/81, rots. G–G dorse; E 364/86, rots. G–G dorse; *CPR 1452–1461*, p. 76; E 101/55/3; E 405/57, m. 7d; Oppenheim, *Naval Accounts and Inventories of the Reign of Henry VII*, pp. 36, 50, 58–9, 69, 73, 194–5, 204–5, 216–17, 247, 261; E 315/317, f. 15r.

4

The Calais Garrison

In 1347 Edward III captured Calais after a siege of almost a year. The town, together with a small enclave of territory around it known as the Pale of Calais, was to remain in English hands for over 200 years, until its recapture by the French in 1558 (see Map 1). It was a highly prized possession for successive kings of England due to its strategic location in the Strait of Dover, which meant that English armies could be transported to and from northern France with relative ease. Yet it also meant that the territory had to be protected by a large and well-equipped garrison. The size of this force varied slightly over time with a muster roll of 1466 listing the names of 527 men serving in the town of Calais, Calais Castle and Rysbank Tower, whereas a later one, from 1502, records 551 soldiers.[1] In wartime reinforcements were sent to bolster the defence of the territory in special companies known as crews.[2] During the Hundred Years' War the Pale of Calais was often threatened with attack, and on occasion, most notably in 1436, was besieged. Later, in the Wars of the Roses, control of the territory was contested between the Lancastrians and Yorkists. This meant that the garrison had to be supplied with substantial quantities of weaponry including guns.

Fortunately, a large quantity of detailed information exists which records the numbers, types and locations of ordnance in the Pale of Calais. This material is available because the Calais garrison was financed by the English exchequer, which necessitated the creation of comprehensive financial accounts and inventories for the purposes of auditing. The most useful source is provided by accounts of the treasurer of Calais. The treasurer was the official responsible for the receipt and expenditure of money; the latter included the payment of the wages of the garrison and the purchase of military equipment. From 1375 to 1485 the accounts of the treasurer were enrolled in the foreign account rolls of the Exchequer and survive as almost a complete series over a 110-year period.[3] These documents exist in a

1 TNA, E 101/55/23; BL, Add. MS, ff. 58–65v; David Grummitt, *The Calais Garrison: War and Military Service in England, 1436–1558* (Woodbridge: Boydell Press, 2008), p. 46.

2 Ibid., p. 56.

3 There is a short gap from 1466 to 1468.

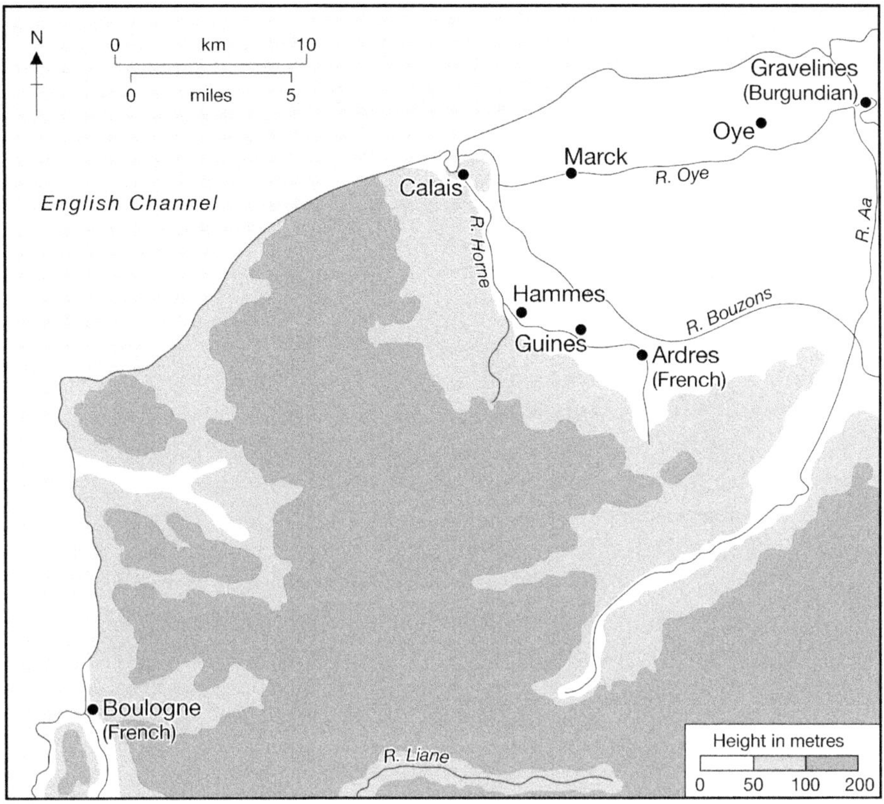

Map 1 The Marches or Pale of Calais in the fifteenth century.
Map drawn by Cath D'Alton.

standard annual format listing the purpose and dates of the accounting period, the receipt of money, equipment remaining from the previous accounting period, the purchase of equipment, the expenditure or sales of equipment and finally the equipment remaining in the account. This means that the quantity of gunpowder, ammunition and guns available to the garrison of Calais can be traced over an exceptionally long period, which allows long-term changes in the use of artillery to be traced in detail.

To a certain extent this information can be supplemented by the extant victualler accounts. The victualler of Calais was the official responsible for the supply of weaponry, other materials and provisions, whose accounts survive intermittently from 1436 to 1485.[4] These documents are similar to the accounts

[4] The office of victualler was abolished in 1492, Grummitt, *The Calais Garrison*, p. 118.

of the treasurers but contain additional details such as where guns were acquired from and how equipment was expended by the garrison. From 1472 to 1488 the victualler of Calais, William Rosse, also had a special account of artillery which was not recorded in the victualler and treasurer accounts.[5] These guns were stored in Calais but appear to have been used for expeditions further afield, such as in 1475 and 1481, rather than for the defence of the territory. Another useful source is a number of 'views' or surveys of artillery in the different fortifications at Calais which survive intermittently from 1474 to 1486 amongst the rough working papers of William Rosse (see Table 9). These give the exact locations of guns and therefore show how artillery was deployed defensively in this period. Finally, some of the accounts of works carried out at Calais survive which provide information on the construction of artillery fortifications.

Only relatively limited evidence exists for guns at Calais after 1485. Early in the reign of Henry VII, reforms were carried out to the financial administration of the Pale of Calais with the responsibility for the supply of victuals and weaponry transferred to royal commissioners.[6] None of these accounts survive for the reign of Henry VII which means that there is very little extant documentation for ordnance at Calais after 1488.[7] Despite this limitation, a large quantity of information survives for artillery at the Pale of Calais for the fifteenth century, which provides an opportunity to identify the key periods of change in the progression of the technology. This chapter will examine how and why the use of guns changed over the period.

CHRONOLOGICAL DEVELOPMENT AND NUMBERS OF GUNS OVER TIME

Gunpowder artillery played an important role in the defence of Calais throughout its time under English control. Edward III used at least ten guns during the siege of Calais and they appear to have been put into the garrison's arsenal soon after its capture in 1347.[8] The earliest evidence of their presence is provided by the receiver of victuals account for 1349–51, which includes the receipt of 14lbs saltpetre and 6lbs sulphur.[9] Later in 1355–7 three guns together with 192lbs saltpetre and 242lbs

5 TNA, E 101/55/5; E 364/119/36, rot. C; David Grummitt, 'The Defence of Calais and the Development of Gunpowder Weaponry in England in the Late Fifteenth Century', *War in History*, 7 (2000), 253–72 (p. 263).

6 Grummitt, *The Calais Garrison*, p. 192.

7 Ibid., p. 119.

8 T. F. Tout, 'Firearms in England in the Fourteenth Century', *The English Historical Review*, 26 (1911), 666–702 (pp. 671, 688–690).

9 TNA, E 101/168/1.

of sulphur were received, most probably from England.[10] Limited activity occurred in the 1360s, with two large guns acquired in 1363–5 and 20lbs of gunpowder allocated for the town of Ardres in March 1369.[11] The resumption of war with France later that year led to an increase in the quantity of ordnance at Calais. This measure was motivated by the rapid French conquest of the neighbouring county of Ponthieu and fears that the Pale was similarly vulnerable to attack.[12] The existing stock of four large guns was augmented by the purchase of fifteen guns, together with gunpowder, saltpetre and sulphur acquired from Bruges in Flanders between 1369 to 1371.[13] The following three years saw a further significant increase with the acquisition of thirty guns and 684lbs of saltpetre.[14] These measures were necessary because Philip the Bold, duke of Burgundy, attacked the Pale of Calais with a large army and was able to capture the castles of Ardres and Audquicq in 1377.[15]

By the beginning of the reign of Richard II, the arsenal at Calais consisted of forty guns, in addition to 1,377lbs of gunpowder and 2,014lbs saltpetre.[16] Over the following five years this was augmented with the purchase of an additional fourteen guns.[17] The greatest increase in artillery occurred in the years 1384 to 1387 when the number of guns at Calais almost doubled from fifty-four to 104. This was undoubtedly motivated by the very real threat of attack posed by the French in these years.[18] After 1387 there were no further major purchases of ordnance for the remainder of the century. From 1393 to 1394, however, there was a significant increase in the number of munitions with the stockpile of gunpowder almost tripling from 1,304lbs to 3,800lbs, and saltpetre seeing a similar increase from 1,120lbs to 2,844lbs, sulphur from 620lbs to 2,014lbs and gunstones from 3,500 to 5,067.[19] It is unclear why these purchases were made as England and France were involved in peace negotiations in this period.[20] Therefore small numbers of guns were initially used by the garrison of Calais in the mid-fourteenth century which began to increase with the renewed outbreak of war in 1369. The reign of Richard II saw the most substantial change in the quantity of artillery with the numbers of guns almost doubling during the invasion threat years of 1384 to 1387. By the

10 TNA, E 101/173/3.

11 TNA, E 101/177/6; E 101/178/14.

12 Jonathan Sumption, *The Hundred Years War: Volume III, Divided Houses* (London: Faber, 2009), pp. 35–6.

13 TNA, E 101/170/15, ff. 3r, 4v, 12v, 13v, 15r–15v, 19r–23r.

14 TNA, E 101/179/6.

15 Sumption, *The Hundred Years War: Volume III*, pp. 291–4.

16 TNA, E 364/14, rots. I–I dorse.

17 TNA, E 364/19, rot. C dorse.

18 Sumption, *The Hundred Years War: Volume III*, pp. 584–8.

19 TNA, E 364/28, rots. B dorse, F.

20 Sumption, *The Hundred Years War: Volume III*, pp. 804–11.

beginning of the fifteenth century guns already played an important role in the defence of the territory.

The numbers of guns at Calais had declined slightly to eighty-five by the time of the accession of Henry IV in 1399.[21] Fears of a French attack to support rebels in England resulted in measures to strengthen the garrison being taken with the purchase of eight guns in 1399–1401 together with large quantities of saltpetre and sulphur.[22] In the following two years continuing tensions with France led to the acquisition of a further eight guns with 1,548lbs of saltpetre, although large quantities of sulphur and gunstones were sold.[23] Despite these measures, one of the king's commissioners at Calais wrote to the mayor of London requesting assistance in guarding the English Channel on 14 October 1404. It was believed that the Pale was in danger of attack and had inadequate supplies of men, artillery and ammunition for its defence.[24] In May of the subsequent year an attack was indeed launched upon Marck Castle in the Pale of Calais. The French were able to capture the town and were in the process of besieging the castle when they were defeated by a relief force sent from Calais. The English force then unsuccessfully besieged Ardres using artillery.[25] In the subsequent year Calais was threatened by a large and well equipped army led by John the Fearless, duke of Burgundy. The English responded by launching a series of attacks on the nearby castles of Columbrine in August and Ballingham in September. In the event the French did not besiege Calais, but they were successful, in starting a fire at the castle of Oye in October, which resulted in the destruction of eight guns, 898lbs of gunpowder and 670 gunstones.[26]

Calais was therefore the scene of much fighting over these two years as is revealed by a list of equipment expended in the engagements at Marck, Ardres, Columbrine and Ballingham castles, which included four guns, 1,126 gunstones, 802lbs gunpowder, 4,197lbs saltpetre and 2,720lbs sulphur. This prompted the delivery of additional supplies of artillery and gunpowder to the territory to replace the damaged equipment, which increased the total number of guns to 119 by 1410.[27] Later in 1412, the garrison was engaged in further action against the supporters

21 TNA, E 364/36, rots. E, I.

22 James Hamilton Wylie, *History of England Under Henry the Fourth, Vol. I. 1399–1404* (London: Longmans, Green and Co, 1884), p. 96; TNA, E 364/36, rot. I.

23 As is often the case with these accounts, the identities of the buyers of this equipment are not specified, Wylie, *History of England Under Henry the Fourth, Vol. I. 1399–1404*, pp. 325–7; TNA, E 364/38, rot. D dorse.

24 Wylie, *History of England Under Henry the Fourth, Vol. I. 1399–1404*, pp. 470–1.

25 James Hamilton Wylie, *History of England Under Henry the Fourth, Vol. II. 1405–1406* (London: Longmans, Green and Co, 1894), pp. 90–4; TNA, E 364/44, rot. B.

26 James Hamilton Wylie, *History of England Under Henry the Fourth, Vol. IIII. 1407–1410* (London: Longmans, Green and Co, 1896), pp. 56–61; TNA, E 364/44, rot. A.

27 TNA, E 364/44, rots. A dorse–B.

of John the Fearless, duke of Burgundy, and the French king in alliance with the Orleanists, with the English launching assaults against the castles of Ryngesham and Ballingham. The latter was captured with equipment including eight guns and eight barrels of gunpowder. The French retaliated by attacking the castles of Hammes, Marck, Oye, Sandegate and Guînes. Large quantities of equipment were expended in these actions by the garrison, including three new guns, 4,910lbs of gunpowder, 1,083 gunstones, thirty-one touches for guns and 136 lead shot. In response to this, significant quantities of supplies were purchased including 2,457lbs gunpowder, 4,562lbs sulphur and 1,710lbs saltpetre.[28]

Limited investment in artillery under Henry V occurred before 1417. The king did, however, visit Calais to greet Sigismund of Luxembourg in 1416.[29] It appears that as part of the celebrations to welcome Sigismund the guns of the town were fired as 1,109lbs gunpowder, 1,550lbs saltpetre, 462lbs sulphur, and 343 gunstones were expended from the accounts in 1415–17. This included one large gun that was said to have ruptured and broken into many pieces in the presence of the king himself. The gun had recently been constructed by Richard Foriet, a gunner of the king at Calais.[30] The weaponry of the garrison was later increased to 132 guns through the purchase of twenty-eight guns, 4,474 gunstones and 750lbs of gunpowder in the period 1417–21.[31] It appears that these purchases were motivated by the threat posed by the supporters of the Dauphin in Picardy in 1421, with Henry crossing to Calais with an army in the same year.[32] The focus of the war soon shifted elsewhere, however, with the treasurer, Richard Buckland, being instructed to deliver weaponry to Thomas Luddington, clerk of the ordnance, which included two large fowlers, 1,118lbs of gunpowder and 120 gunstones on 16 July 1422.[33]

For much of the 1420s, the alliance with Philip the Good, duke of Burgundy, and English victories in France meant that Calais faced little danger of attack. Therefore relatively little was spent on artillery, with the number of guns decreasing very slightly to 129 by 1429.[34] This situation was changed by the lifting of the siege of Orléans by the French later that year and subsequent English defeats, which made the territory more vulnerable.[35] In 1430, the young Henry VI crossed to Calais with

28 James Hamilton Wylie, *History of England Under Henry the Fourth, Vol. IV. 1411–1413* (London: Longmans, Green and Co, 1898), pp. 72–3; TNA, E 364/65, rot. A dorse.

29 James Hamilton Wylie, *The Reign of Henry the Fifth, Vol. III. 1415–1422* (Cambridge: Cambridge University Press, 1929), p. 23.

30 TNA, E 364/69, rot. L dorse.

31 TNA, E 364/57, rot. A.

32 Wylie, *The Reign of Henry the Fifth, Vol. III. 1415–1422*, p. 320.

33 TNA, E 159/199, brevia directa baronibus, Michaelmas, rot. 26.

34 TNA, E 364/65, rot. D.

35 Ralph A. Griffiths, *The Reign of King Henry VI: The Exercise of Royal Authority, 1422–1461* (London: Benn, 1981), p. 188.

a large army for his coronation in France, with three large guns constructed in the town for the expedition.[36] Over the following two years the stock of weaponry was augmented through the purchase of nine guns.[37] In July 1436 the Pale was besieged by a large Burgundian army which was well equipped with ordnance. English preparations to counter this threat included instructions issued by the king to the Exchequer in March of the same year, to purvey twenty-eight guns, 500 gunstones and three barrels of gunpowder. In the town itself, 6,600lbs of gunpowder was purchased or mixed by gunners, with one large gun called *London* and two large fowlers supplied by Alexander Mason, in addition to the purchase of a further three guns. Despite these measures, the Burgundians were able to rapidly overrun most of the outlying castles in the Pale, such as Guînes, Hammes, Oye, Sandegate and Ballingham. The town of Calais itself, however, was able to successfully resist the subsequent siege, in part due to the effective use of its artillery.[38]

It is clear that large quantities of equipment were exhausted by the garrison in resisting the Burgundian attacks. The accounts of the treasurer Richard Buckland for February 1434 to July 1436 show a significant decrease in the number of guns at Calais from 137 to ninety-five, with fifty-four guns expended, in addition to 4,397 gunstones, 7,111lbs gunpowder, 5,911 ½lbs of saltpetre, 2,951lbs of sulphur and 144 lead shot. Another set of accounts for his successor, Robert Whittingham, covers some of the same period, February 1436 to February 1437. A further quantity of equipment including 678lbs of gunpowder, 100 gunstones and two fowlers was stated as expended in the siege. Whittingham also acquired twelve long guns by purchase, in additional to six fowlers provided by William Soper, keeper of the king's ships, together with a large gun captured from the Burgundians. In the period 1436–9, the number of guns at Calais was increased to 119. Furthermore, large quantities of equipment were recovered after the siege, including thirty-seven gunstones from Marck Castle, 172 gunstones from Guînes Castle and 1,051 gunstones from the area outside Calais. In the 1440s the threat to Calais receded which led to its arsenal being neglected. Few purchases for artillery were made in these years and the stock of gunpowder gradually dwindled to 828½lbs by 1446.[39]

36 Anne Curry, 'The "Coronation Expedition" and Henry VI's Court in France, 1430 to 1432', in *The Lancastrian Court*, ed. Jenny Stratford (Donington: Shaun Tyas, 2003), pp. 29–52 (pp. 29–31); E 364/69, rot. Q dorse.

37 TNA, E 364/66, rot. C.

38 Grummitt, *The Calais Garrison*, pp. 20–31; Robert D. Smith and Kelly DeVries, *The Artillery of the Dukes of Burgundy 1363–1477* (Woodbridge: Boydell Press, 2005), pp. 110–16; TNA, E 404/52, nos. 219, 225; E 403/723, m. 11; E 403/725, m. 7; E 364/72, rot. E.

39 TNA, E 364/72, rots. E, M dorse; E 364/74, rots. A dorse–B; E 364/87, rot. G dorse.

French victories in Normandy in 1449–50, however, meant that Calais was once again endangered.[40] Its defence was prioritised by the government, with Gilbert Parr, keeper of the artillery in the Tower of London, ordered to deliver 1,772lbs of saltpetre to Robert Manfield, victualler of Calais, in February 1450. In the following month, Manfield was instructed to 'make asmany gonnes artillaries and other habilements of werre as shall be thought necessarie for the defence and saufgarde of the said towne and marches to be chaunged and newe fourmed' to be paid for by the Exchequer. This led to the purchase of twenty-three guns, including one serpentine, together with 10,246lbs of saltpetre, 1,921lbs gunpowder and 3,300 tampons. An additional fourteen guns were given by Lord Sudeley, a former lieutenant of Calais, including one large gun called *Le Crowne* and a large fowler called *Gloucester*. Herman Donker, the king's gunner at Calais, provided a further thirty-four guns forged from the iron of the twelve long guns, as well as 1,000lbs of gunpowder. This meant that the total number of guns at Calais rose to 160 in 1451, even though eleven guns were also expended this year.[41]

The investment in artillery was not maintained throughout the 1450s with the number of guns falling to 139 by 1460.[42] The stock of gunpowder at Calais continued to decline through the decade, with only 28½lbs available in 1460. By contrast, large quantities of saltpetre continued to be kept at Calais, 3,149½lbs in 1451, 11,279½lbs in 1454, 6,044½lbs in 1457 and 11,390½lbs in 1458.[43] It appears that gunpowder was now only made from its constituent components when required by the gunners of the garrison. This was carried out on a regular basis in this period, with 3,987lbs of gunpowder being expended from 1451 to 1459 in various unspecified engagements against enemies of the king.[44] Later in 1460, the Calais garrison was divided between the supporters of Richard Neville, earl of Warwick, in Calais town and the Lancastrians led by Edmund Beaufort, duke of Somerset, in the castles of Guînes and Hammes. The former expended 356 gunstones, 240lbs of gunpowder and 517 tampons in various skirmishes and sieges. In the following year, Hammes Castle was besieged by the rest of the Calais garrison in March and then again in September and October. The castle was eventually surrendered on 24

40 Griffiths, *The Reign of King Henry VI*, pp. 522–8; Grummitt, *The Calais Garrison*, p. 124; TNA, E 404/66, no. 86.

41 TNA, E 159/227, brevia directa baronibus, Trinity, rot. 3; PSO 1/17, nos. 879, 899; E 364/87, rots. G dorse–H; E 28/81, no. 20.

42 Note that the number of bronze guns was revised down by ten in 1453 without any explanation, TNA, E 364/89, rot. K; E 364/96, rot. B.

43 TNA, E 364/89, rot. J; E 364/90, rot. K dorse; E 364/92, rot. C dorse; E 364/94, rot. A dorse.

44 TNA, E 364/87, rot. H; E 364/89, rots. J–J dorse; E 364/90, rot. B; E 364/91, rots. B–B dorse, C; E 364/93, rots. D–D dorse; E 364/95, rots. A, A dorse; E 364/95, rots. E–E dorse.

October, but the attackers were forced to expend four guns, 3,621 tampons, 1,742 gunstones and 1,748lbs of gunpowder.[45]

Calais, despite the important role it played in the Yorkist victory in 1461, saw relatively limited investment in artillery for much of the first reign of Edward IV (1461–70) with few guns being purchased, although 8,981¾lbs of saltpetre and 21,349lbs of sulphur were acquired in 1462–3.[46] This appears to have led to a deterioration in the condition of the ordnance in the Pale, as the king wrote to the captain of the garrison, Richard Neville, earl of Warwick, and other officials on 20 August 1465 that 'we be duely enformed that of oure stuffe of artillarie abilements of werr…some therof be hold and feble'. They were therefore directed to 'suffisauntly repayer amende and translate all the forseid stuffe of artillarie', which led in the following year to the construction of one large fowler, one serpentine and an organ gun with thirty-nine organs, as well as a gun chamber that was forged out of one of the two remaining parts of the *London*, which had been broken in the siege of 1436.[47] Later in 1468 a marriage treaty was concluded between Charles the Bold, duke of Burgundy, and Margaret of York, Edward IV's sister, which was also accompanied by a commercial treaty and military alliance between the two rulers. This resulted in the purchase of sixteen guns from Flanders, including four serpentines, as well as twenty-one handguns; the first time they are recorded in the Calais accounts.[48] Edward's overthrow by his cousin, Richard Neville, earl of Warwick in 1470, however, led to conflict between the Calais garrison and the Burgundians. Three separate incursions were carried out by the Burgundians into the Pale, with the English also launching raids into Picardy. The victualler accounts record the defensive expenditure of 886lbs of gunpowder, 207 gunstones and 188 lead shot, with a further 297lbs of gunpowder, 133 gunstones and forty-two lead shot used up in offensive warfare. Additional weapons and supplies were also purchased by the garrison at this time, including twenty-four handguns, seventeen hackbuts, 1,206 gunstones, 6,121lbs of saltpetre and 18,000 tampons.[49]

45 Note that the victualler accounts only record the expenditure of three guns at the siege of Hammes Castle, but in this instance less information is recorded than in the treasurer accounts, so it appears to be inaccurate, Grummitt, *The Calais Garrison*, pp. 11–12; TNA, E 364/95, rot. E dorse; E 364/96, rots. B–B dorse; E 101/195/14, f. 10r.

46 There is no evidence from the accounts that guns were taken from Calais by the Yorkists for the campaigns of 1460 and 1461. TNA, E 364/97, rots. G–G dorse.

47 TNA, E 159/247, recorda, Trinity, rot. 7; E 364/101, rot. I dorse.

48 Note that the commercial treaty explicitly excluded the sale of arms and armour between England and Burgundy but in practice this was ignored, Richard Vaughan, *Charles the Bold: The Last Valois Duke of Burgundy* (Woodbridge: Boydell Press, 2002), pp. 60–1; Grummitt, 'The Defence of Calais', p. 261; TNA, E 364/103, rots. G–G dorse.

49 John Riley Rainey Jr, 'The Defence of Calais, 1436–1477' (unpublished Ph.D. thesis, Rutgers University, 1987), p. 98; TNA, E 364/105, rots. E–E dorse.

After Edward IV's return to power in 1471, the territory was again attacked by French soldiers and English rebels, the latter most likely in the service of the fugitive Lancastrian noble, John de Vere, earl of Oxford. From 1471–2 a total of 4,555lbs of gunpowder, 19,000lbs of saltpetre, 5,025lbs of sulphur and five guns were expended from the accounts. The following year, a further seven guns along with 2,469lbs of gunpowder and 761 gunstones were also consumed for the same purpose. Measures to improve the security of Calais resulted in a renewed investment in the weaponry of the garrison from 1471–3, which increased its number of guns to 156 and its number of handguns to 285. This included the purchase from Flanders of twenty guns, 11,500 gunstones, 106 balls of iron called iron gunstones and 223 handguns.[50] From 1472 the victualler of Calais, William Rosse, also had a special account of artillery which was stored in the Pale. This ordnance appears to have been used primarily for expeditions, such as in 1475 and 1481–2, and was not included in the regular accounts.[51] Additional purchases of artillery were made by Richard Whetehill, lieutenant of Guînes Castle in 1472. This consisted of thirty-eight guns, made up of seventeen serpentines, eleven fowlers, one pot gun and nine cart guns in three carts.[52]

Edward IV's 1475 expedition to France used Calais as a base and appears to have drawn upon large numbers of guns from the garrison's arsenal.[53] Regular expenditure of gunpowder occurred for the remainder of the decade but it appears that the garrison was only engaged in minor military actions. In 1479–80, twenty-one guns were recorded as having been expended but only the comparatively small amount of 245lbs of gunpowder and 280 tampons, which suggests that the equipment may have been sold.[54] After this date the quantities of recorded equipment remain largely unchanged until the treasurer accounts finish in 1485, with the number of guns remaining consistently at 183 and handguns at 316.[55] The uniformity of these accounts is misleading, however, as is revealed by the evidence from the views of Calais taken in 1481 and Rosse's enrolled accounts of 1488, which demonstrates that the number of guns located there did fluctuate in this period. The only complete set of views exists for the year 1481 and records that 220 guns and 133 handguns were stored in Calais Town, Calais Castle, Rysbank Tower,

50 TNA, E 364/106, rots. I–I dorse; E 364/107, rots. J–J dorse.

51 Rosse's account was submitted in 1488, TNA, E 364/119/36, rots. A, B.

52 Note these guns were kept at Guînes Castle and were only recorded in the treasurer's account from 1478 onwards, therefore the increase from 156 to 204 guns actually occurred six year earlier in 1472, TNA, E 159/249, recorda, Trinity, rots. 12d–13; E 364/112, rot. C; Grummitt, 'The Defence of Calais', p. 266.

53 See below. By contrast the treasurer and victualler's accounts suggest that none of the garrison's guns were used in the campaign, TNA, E 364/109, rot. D; E 101/198/16.

54 TNA, E 364/114, rot. D dorse.

55 TNA, E 364/115, rot. A; E 364/116, rot. D; E 364/117, rot. B; E 364/118, rot. C; E 364/119/1, rot. A.

FIGURE 2 Total number of guns at Calais 1375–1485.

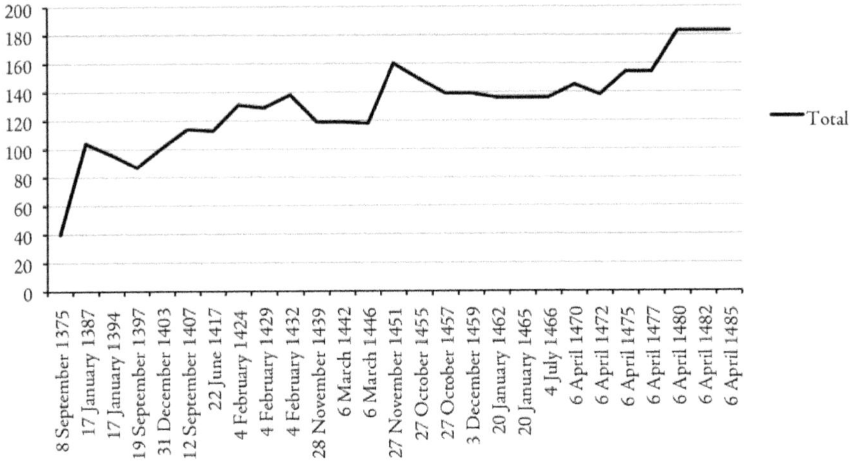

Source: Information derived from the enrolled accounts of the treasurer. TNA, E 364.

Guînes Castle and Hammes Castle.[56] It is likely that most of these guns were part of the regular complement of the garrison's arsenal. The rest of the artillery is likely to have been made up from the remnants of the equipment from Rosse's account, with at least seven large guns, twenty-six serpentines, twenty-four hand culverins and 111 hackbuts being transported from Calais to Scotland in the same year.[57]

Most, if not all, of the ordnance was returned to Calais after the expedition, with Rosse's inventory for 6 April 1485 revealing that 337 guns and 529 handguns were located in the Pale, including the 183 guns and 244 handguns recorded in the treasurer's account for the same year.[58] In addition to this, a further 141 guns and 213 handguns are listed which comprised Rosse's separate stock of artillery for expeditions. The last section of the inventory consists of thirteen large carted guns which correspond to the ordnance delivered by John Sturgeon, master of the ordnance, to William Rosse in 1475 for the expedition to France.[59] In the same year, large quantities of equipment were purchased for the garrison, including 15,444lbs of serpentine powder, 4,000lbs of hackbut powder, 3,000lbs of touch powder, 2,000lbs of bombard powder, eighty-two hackbuts, seven handguns and

56 TNA, E 101/198/13, ff. 76r–85v; 90r–90v.

57 Ibid.., ff. 6r, 8r.

58 TNA, E 364/119/36, rot. B.

59 TNA, E 101/55/7.

FIGURE 3 Total quantities of handguns at Calais 1469–85.

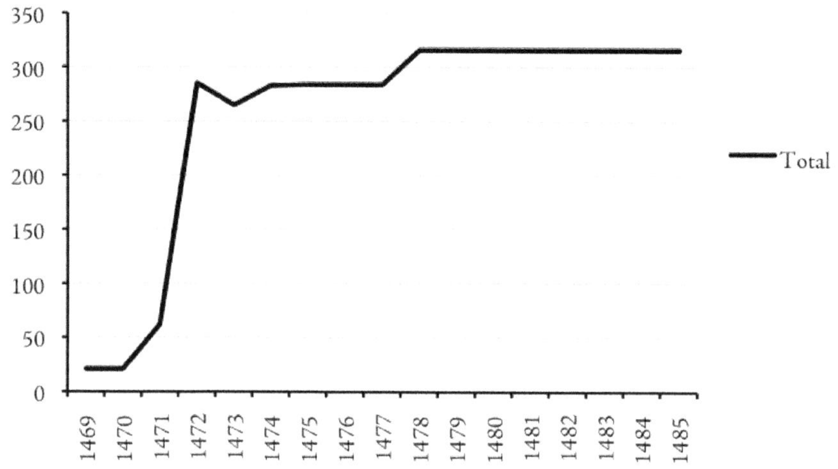

Source: Information derived from the enrolled accounts of the treasurer, TNA, E 364.

FIGURE 4 Numbers of guns and handguns at Calais according to the views of 1481 and 1485, and the inventory of 1488.

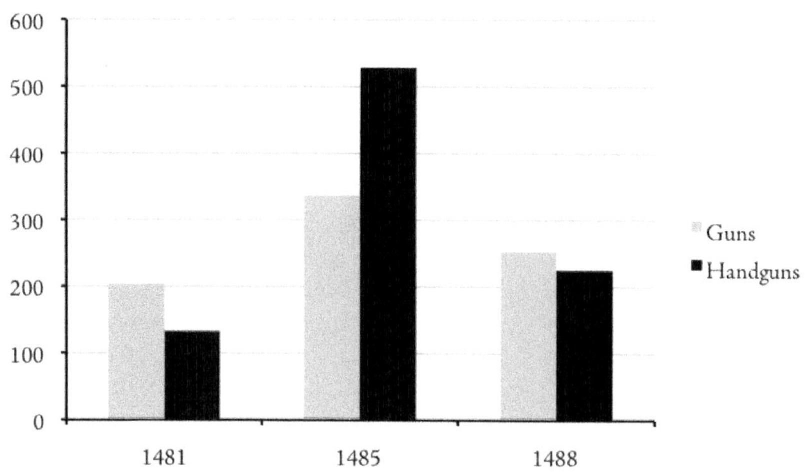

Source: Information derived from TNA, E 101/198/13, E 364/119/336, rots. C, D. Please note that the information given here differs from that of the accounts of the treasurers of Calais.

five serpentines.[60] The ordnance stored at Calais had declined by the inventory of 1488 to 253 guns and 255 handguns.[61] The number of guns in the regular accounts of the garrison stayed mostly unchanged at 183, although the number of handguns decreased to 130. Instead the greatest change occurred to the artillery intended for expeditions, which went down to sixty-four guns and ninety-three handguns, likewise the guns given by John Sturgeon decreased to six. The reason for this marked decrease in the artillery at Calais is not stated but it would appear to be related to changes in the organisation of the ordnance under Henry VII. In 1492, William Rosse was ordered to submit his accounts to the Exchequer with the administration of the Pale thereafter being undertaken by specially appointed royal commissioners. At the same time, Henry VII re-oriented the focus of artillery production from Calais to the Tower of London.[62] It is likely that the process of transferring guns from Calais to the Tower had already occurred by 1488.

Therefore, the number of guns at Calais fluctuated significantly over time from 1375 to 1488. These changes tended to occur in particular periods of tension with France and Burgundy, most noticeably in 1384–7, 1417–21, 1436–9, 1450–1 and 1472–3. Yet for long periods there were few alterations to the quantity of equipment stored at the territory. The scale of investment in artillery at Calais was unique and was unmatched at any other English fortification, which bears testament to the key strategic role of the Pale to successive kings. The importance of Calais was further enhanced by the decision of Edward IV, after his restoration to power in 1471, to use the territory as an arsenal, due to its strategic location and loyalty to the Yorkist regime.

ARTILLERY FORTIFICATIONS

The strategic importance of the Pale meant that it was allocated extensive resources for its defence by the English Exchequer. Therefore the records relating to the town also provide an opportunity to trace the evolution of English artillery fortifications in the fifteenth century. The most useful evidence is provided by the particulars of account of building work carried out at Calais. Only some of these documents have survived, but those that do provide detailed information about the construction and maintenance of the fortifications. This can also be supplemented to a certain extent by incidental references in the accounts of the treasurer. Collectively, these sources demonstrate that important changes in artillery fortifications occurred in the second half of the fifteenth century.

60 TNA, E 364/119/36, rot. C.

61 Ibid., rot. D.

62 TNA, E 159/268, brevia directa baronibus, Michaelmas, rot. 17; Grummitt, 'The Defence of Calais', pp. 267–8.

The town of Calais was first fortified with a stone wall in the thirteenth century which remained largely unchanged after the English conquest in the fourteenth century.[63] There is little evidence for how its defences were modified to adjust to the threat and potential of gunpowder weapons, although it is likely that gunports were added in the late fourteenth century. Similarly, the other fortifications of the Pale, such as Hammes and Guînes castles, had stone defences which apparently were little changed in the fourteenth century. This state of affairs continued into the early fifteenth century, with gunports being added to Guînes Castle in 1403–5 and to Calais Castle in 1409–12.[64] The significant increases in the numbers of guns possessed by the garrison in 1384–7 and 1417–21 were apparently easily accommodated by the traditional defences of the town and satellite castles, therefore.

The military reverses suffered by the English in France in the late 1420s led to the first attempts to adapt extensively the defences of the town to the threat of gunpowder artillery. The treasurer's accounts for the years 1426–8 includes the earliest record of a bulwark at the town, which was constructed on its northern side.[65] Entries in later documents make clear that these structures were designed specifically for the use of guns and were often situated in front of gatehouses.[66] Gunports also continued to be used, with extra ones being added to the walls of Calais between the Beauchamp Tower and Boulogne Gate in 1429–31.[67] These efforts were to intensify when the territory was threatened with attack by the Burgundians in 1436. Two bulwarks were built outside the Boulogne Gate in the period 1434–6.[68] In the following year, another bulwark was constructed outside the northern part of Calais Castle, with a further one made out of earth outside the Boulogne Gate. In addition to this, two gates were built for the latter bulwark, which contained a small lodge, as well as a small bastille which was erected in front of its gates.[69] The town also appeared to have a further two bulwarks made out of wood, with the accounts for 1436–9 referring to one located outside the Milk Gate and another outside the Watergate.[70] Little effort was made to update the other fortifications in the territory before the Burgundian attack, which may have played a part in the relatively easy capture of the castles of Oye, Marck, Ballingham and

63 Grummitt, *The Calais Garrison*, p. 127.

64 TNA, E 364/43, rot. H; E 315/335, f. 11r.

65 TNA, E 364/62, rot. A.

66 For instance, an entry for works carried out from 4–18 October 1449 states that work was carried out on a 'new bulwark for bombards and other ordnance for the safe custody of the town of Calais', TNA, E 101/194/2, m. 89.

67 TNA, E 364/65, rot. D dorse.

68 TNA, E 364/72, rot. E.

69 Ibid., rot. M dorse.

70 TNA, E 364/74, rot. A dorse.

Sandegate, although loops were repaired in Oye Castle in 1436–7.[71] In the three years after the siege, repairs were carried out to bulwarks at Calais, as well as the fixing and framing of a large loop of wood within the walls of Calais for the firing of guns, with a similar loop also being provided for Guînes Castle.[72]

These defences were largely unchanged for much of the 1440s but English defeats at the end of the decade led to new investment in artillery fortifications. A large new bulwark was recorded as being built outside the Boulogne Gate in the treasurer's accounts for 1446–50, with an account of works for 1447–50 recording another bulwark of stone constructed outside Calais Castle. The same account of works also includes a detailed set of payments for the building of a bulwark and gate made of bricks at Guînes Castle which came to a total of £170 1s. 3d. This included the wages of four carpenters who were hired for 108 days, as well as eleven masons who each worked between thirty to ninety-one days, along with thirty-four labourers who each worked between ten to 150 ½ days. Materials used for the bulwark included 4,232 feet of wood and 241,000 bricks.[73] These structures were also provided with windows and loops to allow the firing of guns.[74] Further works to strengthen the southern and eastern sides of the town of Calais were undertaken in the early 1450s. This included one new bulwark constructed outside the Tower of Beauchamp to the east and another bulwark outside the Posterngate to the south in 1450–1; the latter was then later rebuilt in brick in 1453–4.[75] Further works included the building of a bulwark outside Calais Castle in 1451–3.[76]

The reign of Edward IV saw frequent repairs to the fortifications of the town of Calais, including the walls, towers, bulwarks, 'falsebrayes' and murderers.[77] Major works were carried out at Guînes Castle from 1462 to 1474, which has been described as 'one of the earliest fortresses under English control to be adapted to the needs of artillery warfare.' This included the construction of a bulwark made out of brick at the north-west corner of the castle and another bulwark near to the donjon. Repairs were also carried out to the existing defences such as the murderers and 'falsebrayes'.[78] Regular maintenance of the defences in the Pale of Calais continued under Henry VII, but no new artillery works were undertaken.[79] Therefore the adoption of artillery

71 Grummitt, *The Calais Garrison*, pp. 25–6; TNA, E 364/72, rot. M dorse.

72 TNA, E 101/192/17, ff. 21r, 22r, 23v.

73 TNA, E 364/87, rot. G dorse; E 101/194/8, ff. 31r, 37r–38v.

74 TNA, E 101/196/12, m. 3.

75 TNA, E 364/87, rot. H; E 364/91, rot. A dorse.

76 TNA, E 364/89, rot. J dorse.

77 H. M. Colvin, *The History of the King's Works*, volume 1 (London: H. M. S. O., 1963), p. 449.

78 Ibid., p. 453.

79 H. M. Colvin, *The History of the King's Works*, volume 3 (London: H. M. S. O., 1975), p. 339.

fortifications at Calais was mainly a response to military circumstances at specific times. In most instances, new works were undertaken during periods when the stock of ordnance at the territory was increased. This suggests that the guns and artillery fortifications at Calais developed at a similar rate for much of the fifteenth century.

DEPLOYMENT OF GUNS

Artillery played an important role in the defence of fortifications in northern Europe from the fourteenth century. Understanding exactly how it was deployed in the fifteenth century is difficult, however, due to the limited information provided by the main sources of information, financial accounts. This can be seen from the treasurer's accounts and accounts of works at Calais which make frequent unspecific references to the siting of artillery in the circuit of the walls of the town and other fortifications in the Pale, as well as in bulwarks.[80] However, more detailed information survives for the second reign of Edward IV. This is because of the unique survival of a series of 'views' of the artillery deployed in the fortifications of the Pale between 1474 and 1486 (see Table 9). These surveys give detailed information as to the exact locations of guns in the fortifications of the Pale of Calais. This source therefore offers an unparalleled insight into how artillery was deployed defensively in the late fifteenth century.

TABLE 9 Surviving 'views' of Calais.

	1474	1476	1481	1483	1485	1486
Calais (town)		✓	✓			
Calais Castle	✓		✓			✓
Rysbank Tower	✓		✓			
Guínes Castle		✓	✓	✓	✓	
Hammes Castle		✓	✓		✓	✓

Source: Information derived from TNA, E 101/198/13.

These views exist because copies were made by the victualler of Calais, William Rosse, in his rough work book, which were subsequently handed over to the Exchequer in 1492. It has been suggested that these surveys were undertaken either

80 TNA, E 101/192/17, ff. 21r–23v; E 101/194/8, ff. 31r, 37r–38v; E 364/72, rot. M dorse; E 364/78, rot. H; E 364/87, rot. H; E 364/94, rot. A; E 364/119/36, rot. C.

to provide a more comprehensive view of the total ordnance in the Pale of Calais, or alternatively due to Edward IV's decision to invade Scotland in 1481.[81] The dates of the earlier views (1474, 1476 and 1481) certainly suggest that they were linked to the French and Scottish expeditions. As the arsenal at Calais played a key role in the provision of artillery for these campaigns, it appears that Edward IV wanted to have an accurate assessment of how many guns were available to his armies. The low number of guns in 1476, as opposed to 1481, indicates that a significant proportion of the artillery in the Pale had been taken for the French expedition and had not yet been returned. The views of 1483, 1485 and 1486, by contrast, were recorded upon the appointment of new officers to the Calais garrison. This can be seen from the wording of the survey of Hammes Castle for 1486 'A vewe takyn withyn the castell of Hammes the vj daie of Aprill anno primo H vij at the going owte of Sir Thomas Wortley knyght late lieutenaunt of the said castell and at the comyng yn of Sir James Blount knyght now lieutenaunt of the said castell of alman' of stuffe and habiliments of werr that remayn this daie'. It is also likely that other surveys were undertaken but were not recorded in Rosse's book. This is hinted at by instructions issued by Richard III on 28 June 1483 for commissioners 'to take a vewe of the artillerie and habiliments of were of Calaise, Guysnes, and Hammes, and to bring writing therof to the king'.[82]

The town of Calais was the most important location for the deployment of guns in the Pale, as is revealed by the two extant views for the years 1476 and 1481. In 1476, the town had a total of fifty guns, with more than half of them located in three bulwarks, called Boulognegate, Bechyin and Posterngate and smaller numbers allocated to the walls, towers, Laterngate and elsewhere (see Figure 5).[83] The majority of the ordnance was comprised of fowlers and serpentines, together with a smaller number of bombards and trestle guns (see Figure 6). The distribution of these gun types, on the whole, does not appear to follow any clear cut pattern in this year, as can be seen by looking at the three bulwarks. Boulognegate bulwark had a mixture of guns, consisting of three fowlers, four serpentines and one trestle gun, whereas the Posterngate bulwark was solely equipped with nine fowlers, and the Bechyin bulwark was armed with eight serpentines and one fowler. These bulwarks were also supplied with sizeable quantities of gunstones and tampons, which meant that more than ten gunstones and thirty tampons were available for each weapon. No information is provided in this view as to the location of other munitions such as gunpowder or saltpetre.

81 TNA, E 101/198/13, f. 48r; Grummitt, 'The Defence of Calais', pp. 266–8.

82 James Gairdner, ed., *Letters and Papers Illustrative of the Reigns of Richard III and Henry VII*, volume 1 (London: Longman, Green, Longman, and Roberts, 1861–3), p. 15.

83 TNA, E 101/198/13, ff. 63r–64v.

Figure 5 Distribution of guns in Calais Town in 1476, excluding handguns.

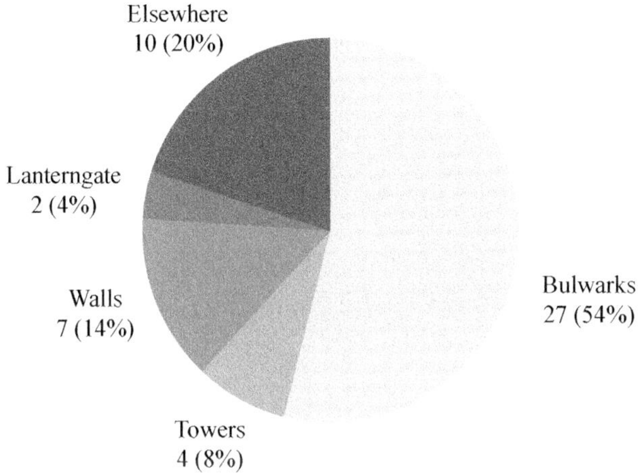

Source: Information derived from TNA, E 101/198/13.

Figure 6 Types of guns in Calais Town in 1476, excluding handguns.

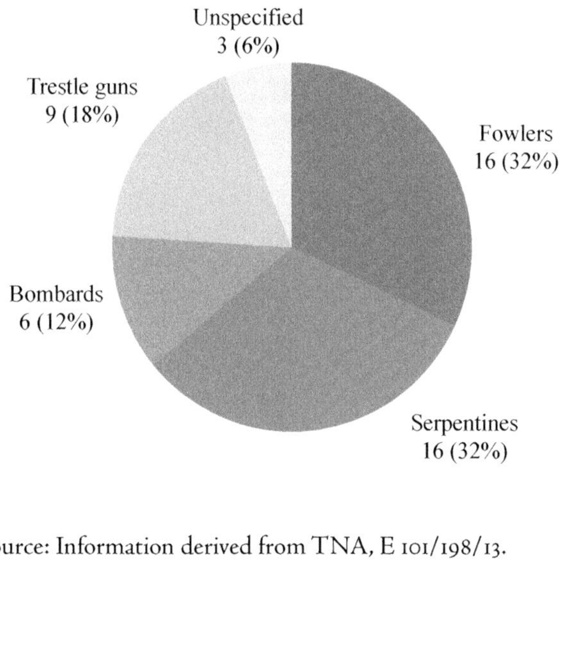

Source: Information derived from TNA, E 101/198/13.

FIGURE 7 Distribution of guns in Calais Town in 1481, excluding handguns.

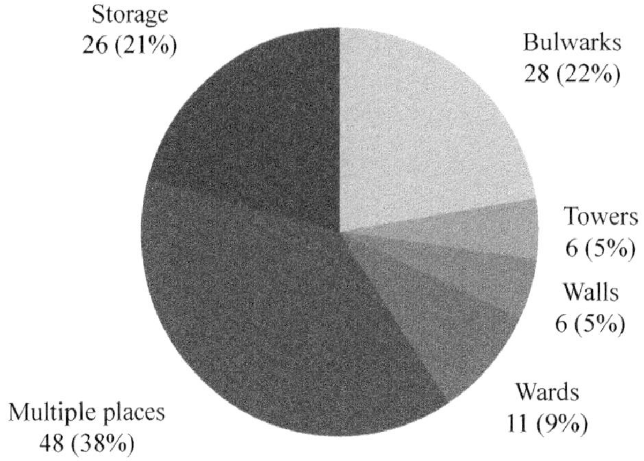

Source: Information derived from TNA, E 101/198/13.

By the time of the 1481 view, the number of guns at the town of Calais had increased from 50 to 125.[84] The quantity of ordnance stored in the bulwarks, walls and towers remained largely unchanged from the previous inventory. A significant difference, however, was the presence of sixty carted guns noted as being in storage or 'multiple places' (see Figure 7). This also affected the proportions of guns at Calais, with the number of serpentines increasing to sixty-four (51 per cent), the number of fowlers to forty-four (35 per cent), and the number of bombard to thirteen (10 per cent) (see Figure 8). The vast majority of the serpentines were carted (80 per cent), whereas much smaller numbers of fowlers and bombards were carted (10 per cent and 30 per cent respectively) (see Figure 9). This demonstrates that serpentines were the preferred form of mobile artillery, which were supplemented by much smaller quantities of bombards and fowlers. These carted guns almost certainly comprised the bulk of William Rosse's special stock of ordnance, which were used as part of the 1475 expedition and had not been returned by the time of the 1476 view. This also seems to have been the case for the town's sixty-one handguns, which were listed as being in the Old Armoury in 1481.[85] Unlike in the previous view, much more detailed information is provided regarding the storage of gunpowder and ammunition in Calais. Two of the towers in the town were described as gunpowder towers which collectively contained forty-eight barrels and one firkin of gunpowder, together with pestles, mortars, rams, staves and 2,150

84 TNA, E 101/198/13, ff. 77r–82v.

85 TNA, E 101/198/13, f. 78r.

Figure 8 Types of guns in Calais Town in 1481, excluding handguns.

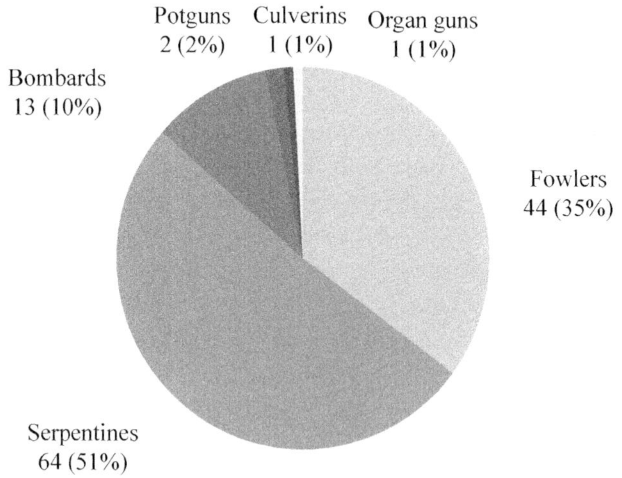

Source: Information derived from TNA, E 101/198/13.

Figure 9 Proportion of carted and uncarted guns in Calais Town in 1481.

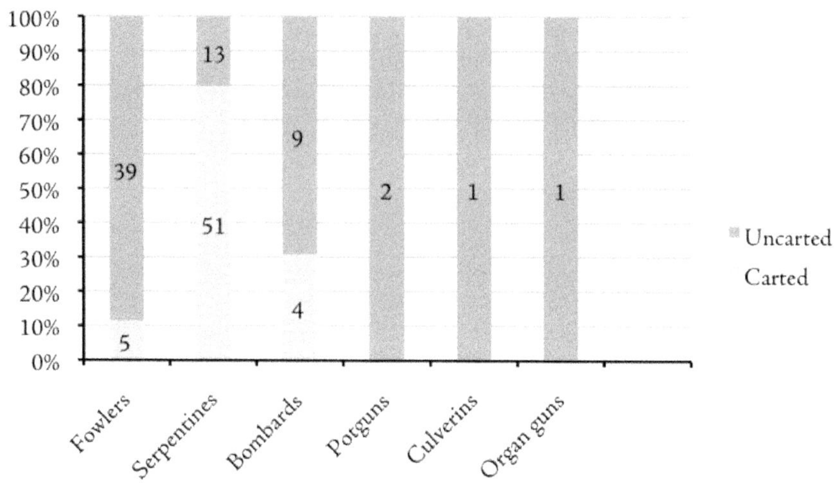

Source: Information derived from TNA, E 101/198/13.

tampons. Almost all the ammunition was stored in the 'House under the Rodr', consisting of 11,834 gunstones, 1,421 lead shot for serpentines, 110 iron shot, 752 lead shot for hackbuts, 1,450 lead shot for culverins, 13,000 tampons and 2 chests of sulphur. Important other locations for the storage of gunpowder and ammunition in the town included the Southwest Tower, the 'House under Mylkgate', the 'bowam' of Beauchamp Tower and the 'Gunpowder House under old Calais'.[86]

Three views survive for Calais Castle for the years 1474, 1481 and 1486, which show that the relatively small number of guns kept there did change over time. In 1474, half of the twelve guns of the castle were deployed in the towers, namely the Watch Tower, South Tower, 'Tower under the Watch Tower' and East Tower, with the rest placed in the 'Balowrt' as well as the Great Hall and Great Chamber.[87] By 1481 the number of guns had declined to nine, with four guns placed in the Watch Tower, Gate Tower and 'Tower under the Watch Tower', with the remainder in the Nursery, Storehouse and Postern Bulwark. In both years, additional equipment was located in the Storehouse, which contained forty-four gunstones, fifty tampons as well as spare gun chambers in 1474, and thirty gunstones, eighteen lead shot and 100 tampons in 1481. The number of guns in the castle had increased to sixteen by 1486, of which seven were located in the 'Balowrt', three in the Donjon, Chapel Tower and But Tower, as well as five in or under the Wool House, 'Psles House' and Nursery. A further gun was stored 'Underneath the But Tower', together with 200lbs of gunpowder, 157 gunstones, 152 lead shot and 230 tampons. Six hackbuts were also recorded as located in the Armoury which contained 132 lead shot for hackbuts.[88] Additional supplies of equipment were delivered to the constable, John Lynche, in the same year, including 400lbs of serpentine powder, four hackbuts, one serpentine and 784 lead shot.[89] Most of the types of guns in 1474 are unspecified, but the other two views make clear that serpentines were preferred for the defence of the castle, which comprised five out of nine guns in 1481 and eleven out of sixteen guns in 1486.

Only two views are extant for Rysbank Tower, which show that its number of guns increased only very slightly from six in 1474 to seven in 1481.[90] The earlier view also contains detailed information on the munitions which were allocated to each piece of artillery. Two serpentines were located 'on the lede' together with two small trestles, seventeen lead shot, 100 new tampons, eighty small gunstones, one firkin of gunpowder and one rest of iron. The 'Lardys House and P'son House' contained fifty tampons together with two guns, the first with twenty gunstones and the second with fifty gunstones. Two cart guns with two carts were stored

86 Ibid., ff. 81r–82r.

87 It is unclear what 'balowrt' means.

88 TNA, E 101/198/13, ff. 42r–42v, 67r–68v, 83r–83v.

89 Ibid., f. 45r.

90 Ibid., ff. 69r–69v, 84r.

Figure 10 Types of guns in Guînes Castle in 1476, excluding handguns.

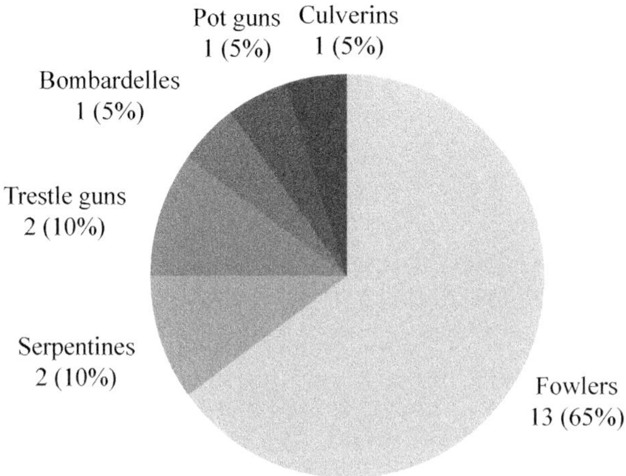

Source: Information derived from TNA, E 101/198/13.

Figure 11 Types of guns in Guînes Castle in 1481, excluding handguns.

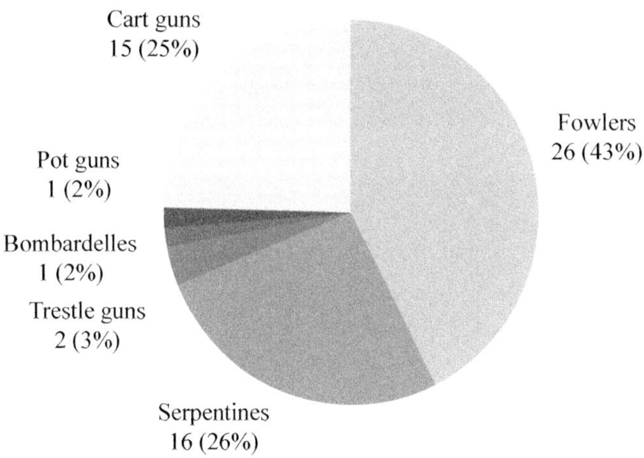

Source: Information derived from TNA, E 101/198/13.

in the Porter's Lodge with fifty gunstones and fifty tampons. The same location also contained twenty-one lead shot and sixty gunstones for serpentines, sixty gunstones for the 'gun in the hall', twenty-four gunstones for the gun in the 'Lardr', sixty gunstones for the gun in the 'P'son House', thirty-three gunstones of diverse sorts and 150 tampons. Twenty lead shot were also recorded as being in the Archery.[91] By the time of the 1481 view, four of the seven guns were listed as being in the 'Botham of the dyke' with the remaining three guns recorded in the 'Landys Bowst and Presor House' and 'About on the lady'.[92] Three hackbuts were also stored in the middle chamber with stampers of iron. Both inventories suggest that most of the ordnance at Rysbank Tower was kept in storage until required for use. This would mean that a speedy mechanism for deploying the artillery must have been in place, as the tower was used for guarding the harbour of Calais and was recorded as providing cover for ships fleeing into the port.[93] The diversity in types of guns, a combination of fowlers, serpentines and cart guns, also indicates that no one type was considered particularly suitable for its defence.

The four surviving views for Guînes Castle, 1476, 1481, 1483 and 1485, demonstrate that it was the most important location in the Pale of Calais for artillery after the town of Calais itself due to its frontier location.[94] In 1476, twelve of the twenty guns at Guînes, including the largest ones, were situated in the Old Bulwark, with a further seven guns on the walls and another in the Archery.[95] The number of guns stored at the castle had risen to sixty-one by the time of the 1481 view, which were simply described as being located on the walls, bulwarks and murderers. The smaller pieces of ordnance, comprising forty-four handguns of different sorts and five small culverins were stored in the Archery. The compilers of the view also recorded large quantities of other equipment stored in diverse locations in the castle, including 3,480 gunstones, 1,500 tampons, 1,870 lead shot and six barrels of saltpetre powder. In addition to this, eleven barrels of gunpowder, five and a half barrels of serpentine powder and three firkins of culverin powder were listed as being in the Gunpowder Tower.[96]

The number of guns at the castle had risen slightly to sixty-six by the time of the 1483 inventory (see Figure 12).[97] Almost half of the guns (thirty) were deployed in the bulwarks, walls and towers, with a further seventeen located in the 'Little Bastlort', braye, gate and murderers. Nineteen of the guns, however, were scattered about in diverse locations in the castle, such as in the chamber wards, the 'Archery

91 Ibid., ff. 69r–69v.

92 Ibid., f. 84r.

93 TNA, E 364/93, rot. D.

94 TNA, E 101/198/13, ff. 2r–4v, 70r–71r, 84v–89v.

95 Ibid., ff. 70r–71r.

96 Ibid., ff. 84v–85v.

97 Ibid., ff. 86r–89v.

FIGURE 12 Distribution of guns in Guînes Castle in 1483, excluding handguns.

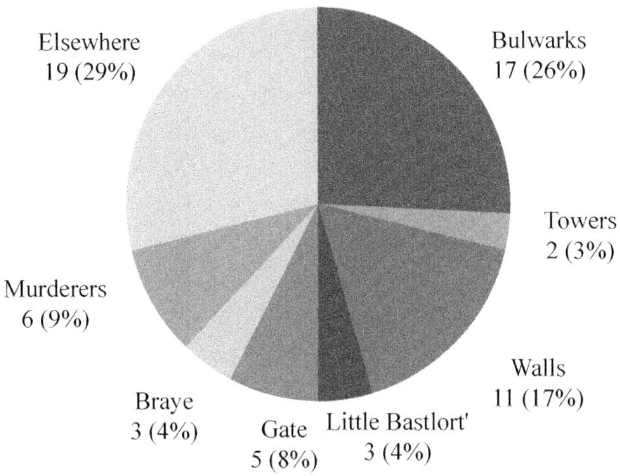

Source: Information derived from TNA, E 101/198/13.

upon the walls', in the 'New garden', by the forge and watch door. In addition to this, thirty-four handguns of different sorts were kept in the Archery, with nine more located in the 'Murderer by the Braye'. This demonstrates that a significant portion of the artillery was not actively deployed in the fortifications. Additional equipment was stored in the Mill House which contained 1,450 gunstones, with the tower of the same storing 1,150 tampons, 1,330 lead shot and three iron shot. Large quantities of gunpowder continued to be kept in the Gunpowder Tower, consisting of five barrels of gunpowder, 131lbs of serpentine powder, 187lbs of culverin powder together with seven barrels of bombard and serpentine powder which weighed 1,989lbs. Further quantities of gunpowder for the castle were purchased later the same year for £234 8s. 10d., made up of 1,682lbs of serpentine powder in fifteen barrels, 534lbs of hackbut powder in two and a half barrels, 187lbs of touch powder in three firkins and 1,250lbs of bombard powder in five barrels.[98]

The 1485 inventory shows that the number of guns at Guînes Castle had slightly declined to sixty-one (see Figure 14).[99] A higher proportion of the ordnance was deployed in the fortifications, however, with twenty-four guns located on or below the walls, as well as sixteen guns in the bulwarks and a further thirteen guns placed in the 'bastots', Braye, Blockhouse and Gate. The remaining seven guns were recorded in the Porter's Lodge, Little Storehouse and Keep, with the New Bulwark also containing

98 Ibid., f. 89v.

99 Ibid., ff. 2r–4v.

forty-three handguns of different sorts. In addition to this, the Gunpowder Tower contained 5,300lbs of gunpowder which were stored in twenty-two barrels and one firkin, as well as two barrels of touch powder, 750 lead shot, three iron shot and 3,200 tampons in the Old Storehouse. The total number of guns at Guînes Castle, therefore, increased greatly from twenty in 1476 to sixty-one in 1481, but stayed at roughly sixty up until 1485. There was also a preference for the larger guns to be placed 'below' the fortifications, for instance with six of the nine guns deployed 'below the walls' in 1485, being large fowlers, long serpentines, pot guns or bombardelles.

FIGURE 13 Types of guns in Guînes Castle in 1483, excluding handguns.[100]

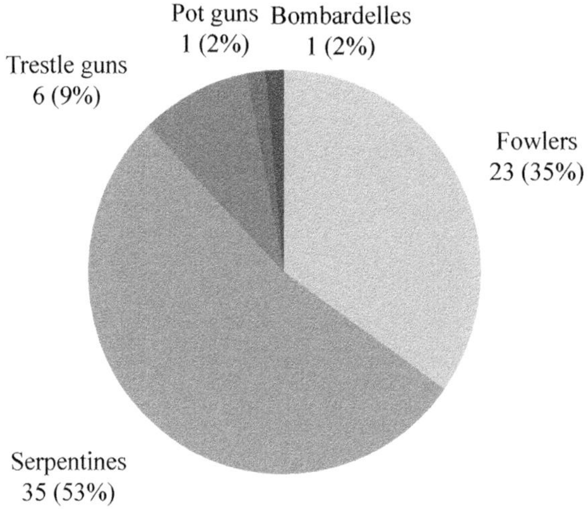

Source: Information derived from TNA, E 101/198/13.

Significant changes occurred to the types of guns deployed at the castle over time (see Figures 10, 11, 13 and 15). In 1476, 65 per cent of the twenty guns were fowlers, a proportion which had fallen to 43 per cent by 1481, despite the number of fowlers doubling to twenty-six. Instead half of the other guns at the castle were made up of cart guns (24 per cent) and serpentines (26 per cent). By 1483, the number of serpentines had more than doubled to thirty-five to make up 53 per cent of the total, with most of the rest made up by twenty-three fowlers (35 per cent) and six trestle guns (9 per cent); proportions which stayed almost exactly the same for the 1485 inventory. This increase in the number of serpentines at Guînes Castle may have been due to an intention to increase the range of the castle's firepower.[101] It may

100 Information derived from TNA, E 101/198/13.

101 See Appendix J for a discussion on serpentines.

also be due, at least in part, to the movement of guns in the reign of Edward IV for his French and Scottish expeditions. In 1472, thirty-eight guns were purchased by Richard Whetehill, lieutenant of Guînes Castle, made up of seventeen serpentines, eleven fowlers, nine cart guns and one pot gun.[102] Yet most of this ordnance was evidently not located in the castle in this year, which only contained twenty guns in total. A similar unexpectedly small number of guns were recorded as located at the town of Calais at the same time. This would indicate that large quantities of artillery were taken by Edward IV for his 1475 campaign, which was not returned to the Calais garrison until after July 1476.[103]

FIGURE 14 Distribution of guns in Guînes Castle in 1485, excluding handguns.[104]

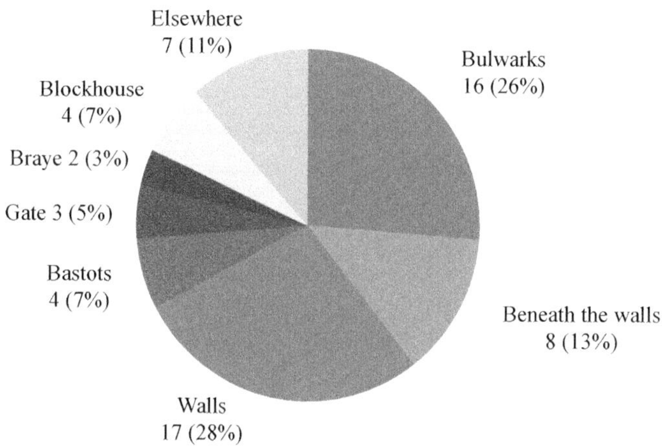

Source: Information derived from TNA, E 101/198/13.

Hammes Castle was another outlying fortification in the Pale of Calais in the reign of Edward IV. Yet the four surviving views of its artillery, for 1476, 1481, 1485 and 1486, show that it was equipped with far fewer guns than Guînes Castle, possibly due to its location.[105] George Ferrers, writing in 1558, described the castle of Hammes as 'of old workmanship, without rampiers or Bulwarks: yet, nevertheless,

102 TNA, E 159/249, recorda, Hilary, rots. 12d–13, I would like to thank Nicholas Kingwell for providing this reference.

103 The view for Guînes Castle was taken in March, with the view for Calais Town taken in July, TNA, E 101/198/13, ff. 63r, 70r.

104 Information derived from TNA, E 101/198/13.

105 Ibid., ff. 48r–49v, 72r–72v, 90r–92v.

by the natural situation thereof … could not easily be approached unto: either with great ordnance for the battery, or else with an army to encamp there, or a siege'.[106] The first two inventories do not provide a breakdown of the location of ordnance at Hammes Castle, but they do reveal that, as with the town of Calais and Guînes Castle, the number of guns increased significantly from six in 1476 to eighteen in 1481, with the latter also including twenty-three handguns.[107] It is difficult to compare the types of guns in both of these views, as the majority of ordnance in 1481 is unspecified. The presence of one carted serpentine and two cart guns, in the latter case, does suggest that a small quantity of mobile artillery had been moved to the castle at some point between 1476 and 1481.

The view of 1485 is of particular interest as it was taken after a siege which had occurred in December of the previous year. John de Vere, earl of Oxford, had been held as a prisoner in the castle since 1474. At the end of 1484 he was able to persuade his jailer, James Blount, lieutenant of Hammes Castle, to release him from captivity and to support the cause of Henry Tudor. The garrison at Hammes Castle was then subsequently besieged by a pro-Ricardian force led by John, Lord Dynham, which was drawn from the rest of the Calais garrison. The castle had been recaptured by January of 1485 when a pardon was granted to the survivors. The view itself was taken in July, six months after the siege, but the relevant section in Rosse's account is prefaced by the title 'after the siege', which suggests that the deployment of guns there was still influenced by the event. In total, twenty-one guns and twenty-three handguns were listed on the walls, towers, murderers, gatehouse and elsewhere (see Figure 16).[108] The most important area for the deployment of the artillery was the murderers with seven guns, which comprised the east and west murderers in the bastion, as well as the south murderer and the 'murderer before the slues'. A further five guns were recorded as lying inside or in front of the Storehouse or 'against the castle'. The remaining nine guns were allocated to the walls, towers and gatehouse.

The distribution of the gun types suggests that certain weapons were deployed in different roles, with the vast majority of guns (76 per cent) consisting of small fowlers or serpentines (see Figure 17). The small fowlers made up the majority of guns deployed in the murderers, gatehouse and 'under the Mycell Tower', nine out of eleven; by contrast serpentines comprised four out of five guns located on the walls and Wardrobe Tower. This would suggest that the serpentines, with their longer barrels, were placed in elevated places to enhance their long range, whereas

[106] Colvin, *The History of the King's Works*, volume 3, pp. 370–1.

[107] TNA, E 101/185/13, ff. 72r–72v, 90r–90v.

[108] TNA, E 101/198/13, ff. 91r–92v; Grummitt, *The Calais Garrison*, p. 14; Charles Ross, *Richard III* (London: Eyre Methuen, 1981), p. 202; Steven Gunn, 'Vere, John de, Thirteenth Earl of Oxford', *Oxford Dictionary of National Biography*, Oxford University Press, 2004; online edition, Jan. 2008 http://www.oxforddnb.com/view/article/28214?docPos=3 [accessed 22 December 2014].

FIGURE 15 Types of guns in Guînes Castle in 1485, excluding handguns.

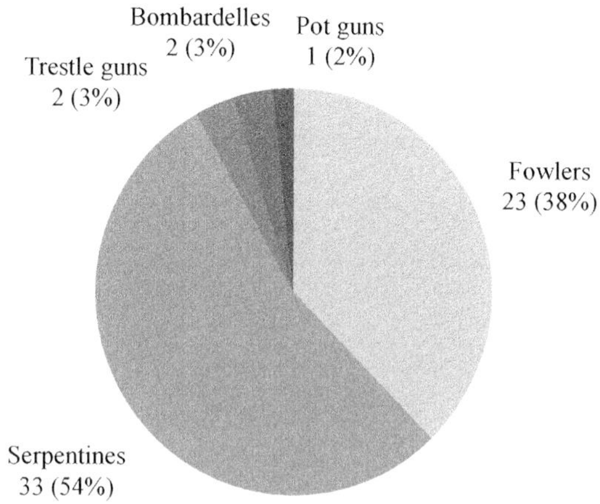

Source: Information derived from TNA, E 101/198/13.

FIGURE 16 Distribution of guns in Hammes Castle in 1485, excluding handguns.

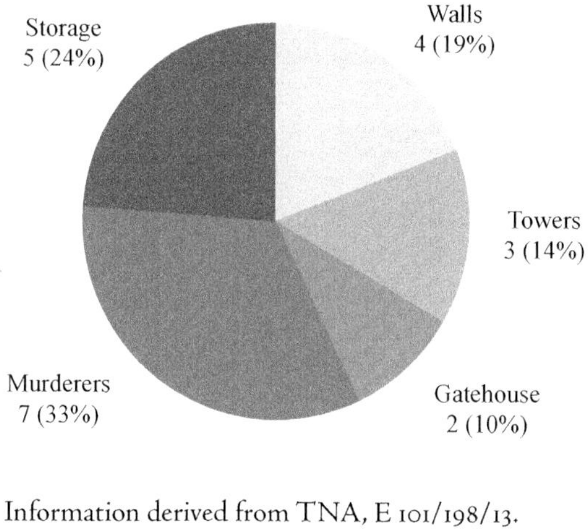

Source: Information derived from TNA, E 101/198/13.

the smaller fowlers were distributed in the most vulnerable parts of the castle, such as the gatehouse, where their short range would be most suitable against attackers. The larger guns, as well as the handguns, were located around the Storehouse, most likely so that they could be deployed as and when necessary. In addition to this, the storehouse contained 1,300 gunstones, sixty lead shot, one and a half barrels of serpentine powder and half a barrel of bombard powder. A 'gunning house' was also used as a location for the storage of one barrel and one tub of sulphur.

FIGURE 17 Types of guns in Hammes Castle in 1485, excluding handguns.

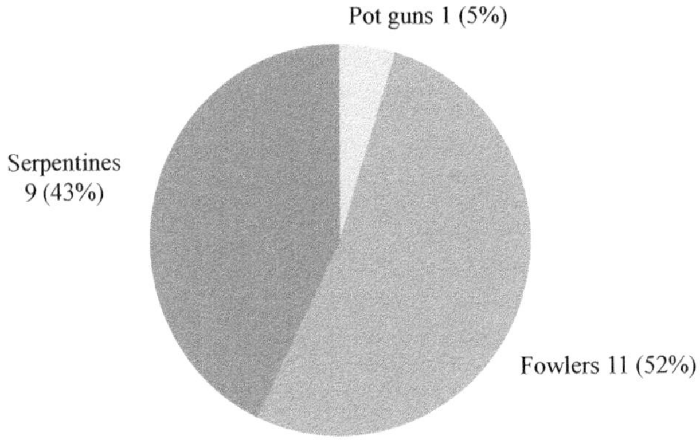

Source: Information derived from TNA, E 101/198/13.

Furthermore, the 1485 view includes information on ordnance which was sent to the attackers during the siege, as well as to strengthen the garrison later in 1485. The besiegers were given the substantial quantity of 9,800lbs of gunpowder, with a further 1,180lbs despatched to Hammes Castle over the following two months. Additional equipment sent to the castle in August included a large fowler and nine hackbuts. By the time of the 1486 view, the number of guns had decreased very slightly to twenty and the number of handguns to sixteen (see Figure 18). A comparison of the two views shows that there was little change in the distribution of the guns in the castle, but that there was a greater diversity of gun types by 1486 (see Figures 16, 17, 18 and 19). The number of 'average' sized serpentines decreased but was partly replaced by other types of guns, such as one trestle gun and one carted serpentine.

The views of the Pale of Calais confirm that bulwarks were a key defensive feature, which were often equipped with guns. Artillery was also deployed in other places, including towers, walls, murderers, bastions and gatehouses, as well as being placed in storage. In many instances, guns were allocated by type, for instance serpentines were often located in elevated positions, such as on towers or bastions, whereas fowlers tended to be placed on walls or murderers. The number and proportion of serpentines also increased greatly from 1474 to 1486, but this is likely to have been due to the movement of mobile artillery during the French expedition.

FIGURE 18 Distribution of guns in Hammes Castle in 1486, excluding handguns.

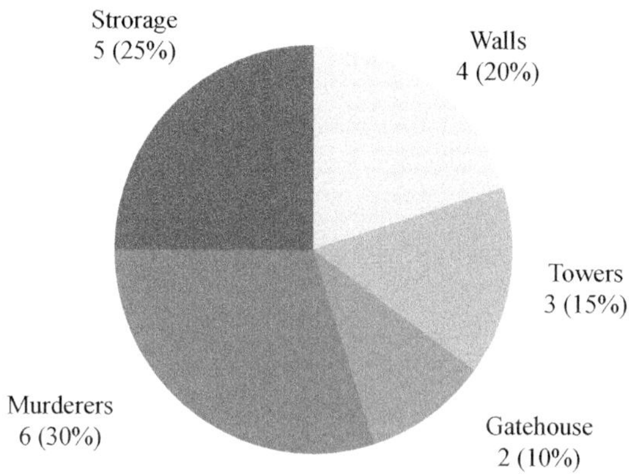

Source: Information derived from TNA, E 101/198/13.

FIGURE 19 Types of guns in Hammes Castle in 1486.

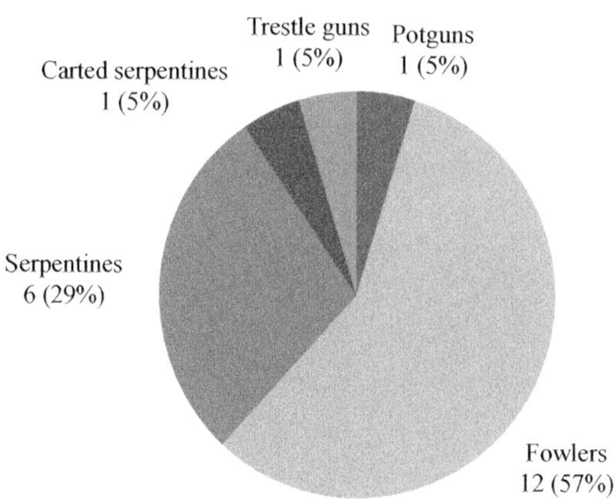

Source: Information derived from TNA, E 101/198/13.

PERSONNEL

The financial documentation for the Calais garrison, as we have already seen, provides plentiful evidence as to the numbers and types of guns stored in the territory. However, there is little information as to the personnel who constructed, maintained and used these weapons. This means that it is difficult to accurately ascertain how many gunners were employed at Calais, their tasks and length of military service. Yet it is possible to broadly trace the activities of the main gunners in the territory over the period using incidental evidence from the accounts of the treasurer of Calais, as they often mention their names in relation to certain tasks such as the making of guns and gunpowder. This information needs to be treated with caution as it is far from comprehensive; for instance, the accounts of the treasurer for 1465–6 simply records that the garrison included Robert Potte, master gunner, and other gunners.[109] It does reveal, however, that the length of service and tasks undertaken by these individuals changed notably over the course of the period.

The earliest recorded gunner at Calais is a Hugh de Rungacto who was described as a 'gunner of the king' in the receiver of victuals' account for 1355–7.[110] Yet it is unclear if a gunner was retained on a permanent basis in the territory, however, as there is no further evidence for one prior to 1375, when repairs to the house of Master William Newlyn are recorded (see Table 10).[111] In the crisis years of 1384–7 the number of guns at Calais increased dramatically, which was matched by the employment of more than five gunners (see Tables 11 and 12). The accounts record that Thomas Gymour was chief gunner at this time, together with his assistant Richard Selby, and Dederico van Ludyk.[112] Other gunners were also employed at the other fortifications at the Pale, including two master gunners at Oye Castle, as well as unclear numbers at Sandegate and Hammes.[113] These gunners, with the exception of Gymour, were hired on a temporary basis due to the fear of a French invasion. This pattern was to continue throughout the fifteenth century, with the small permanent establishment at the territory augmented by extra gunners in crisis years.

109 TNA, E 364/101, rot. H.

110 TNA, E 101/173/3.

111 TNA, E 101/180/5, f. 12v.

112 TNA, E 364/22, rots. H–H dorse.

113 The entry in the issue roll records that ten gunners, crossbowmen, masons, carpenters and armourers were in receipt of wages in Sandegate Castle, with a further thirty such professionals employed at Hammes Castle, TNA, E 364/44, rot. A dorse; E 403/502, m. 4.

Table 10 Main gunners at Calais.

Name	Period of service	Number of years	Note
Hugh de Rungacto	1355–7	2	
Unknown	1357–74	N/A	
William Newlyn	1375–7	2	
Unknown	1378–83	N/A	
Thomas Gymour	1384–90	6	
William Brettevill	1390–1403	13	First served at Calais from 1387–90
William Gerardson	1403–15	12	Possibly succeeded Brettevill in 1404
Godfrey Goykyn	1415–17	2	
Unknown	1417–34	N/A	
Herman Donker	1434–63	29	
Robert Potte	1463–72	9	First served at Calais from 1462–3
Egido van Rasynghin	1472–5	3	First served at Calais from 1467–8
William Donker	1475–1502	27	

Source: Information derived from TNA, E 364, E 101, E 403.

William Bretteville is first recorded as serving in Calais in 1387 and after 1390 had succeeded Gymour as chief gunner. It appears that he left office around 1404, as he was recorded as the former gunner of the king in the accounts of the treasurer for 1403–5, with his replacement being a William Gerardson. A surviving petition to Henry IV, with a response dated 22 April 1404, was made by Bretteville, in which he requested that he have two English assistants born in England to aid and take his craft for the safe keeping of Calais at the king's wages.[114] This is of interest as Bretteville's name suggests that he was of French origin and previous gunners, such as Dederico Van Ludyk, appear to have been from the Low Countries. It is unclear why a preference was expressed for native born Englishmen at this time, but it may have been linked to concerns over the safety of the territory in 1384. The document also includes a note in which the king assented to the bill, so he presumably sent assistant gunners to the town. It is possible that John Barton and William Sim', who

[114] TNA, SC 8/324/E3662.

were employed in the years 1405–7, were sent to the territory by Henry IV for this purpose.[115] In future years there appears to have been little or no attempt to recruit English gunners, as Gerardson was replaced by Godfrey Goykyn in 1415, who is likely to have originated from the Low Countries, and stayed at Calais until 1417.

Table 11 Main gunners at Guînes Castle.

Name	Period of service	Number of years
John Beer	1434–9	5
John Eryom	1436–9	3
John Longer	1450–1	1
Guyutyid Bore	1472–3	1

Source: TNA, E 364, E 159, E 101, E 403.

Thereafter there is no further mention of gunners at Calais until 1430, when three gunners, Giles Thorneton, Alex Mason and Godfrey Goykyn, were paid to go from the town to Flanders to obtain gunpowder for the king's expedition to France.[116] Later, on 16 December 1435, Herman Donker and John Boston were paid for two months' service in the Pale, together with two assistant gunners. These men were sent to reinforce Calais following the defection of the Burgundians earlier that year.[117] At the same time, John Beer is also described as serving in Guînes Castle, the first time a gunner is mentioned there (see Table 11).[118] During the siege of 1436, the establishment at the Pale included Herman Donker, his assistant gunners plus John Terry in Calais, whereas John Beer and John Eryom were assigned to Guînes Castle.[119] Donker continued to be employed at Calais after the siege, with other gunners periodically mentioned by name in periods of crisis, such as in 1450–1 and 1456–7.[120] Robert Potte replaced Donker as the chief gunner in Calais in 1463, a position which he held until 1472; as with his predecessor he was likely to have been Dutch or Flemish. It has been noted that Edward IV recruited many of his

115 TNA, E 364/44, rot. B.
116 TNA, E 364/69, rot. Q dorse.
117 TNA, E 403/721.
118 TNA, E 364/72, rot. D dorse.
119 TNA, E 364/74, rot. A.
120 TNA, E 364/78, rot. H; E 364/87, rot. H; E 364/91, rot. B; E 364/92, rot. C; E 159/249, recorda, Hilary, rot. 8.

gunners from the Low Countries, which almost certainly was motivated by the desire to benefit from their technical expertise.[121]

Potte's successor, Egidio Van Rasynghin, played an important part in the preparations for the 1475 expedition to France. This can be seen from the payments that were made to him for the construction of guns, and later from his role as master gunner of the ordnance company for the campaign.[122] Calais, therefore, not only supplied much of the artillery for the venture, as we have already seen, but also at least part of the personnel. William Donker succeeded Rasynghin in 1475, a position which he was to hold until at least 1502.[123] His time at Calais witnessed an important change in the number of gunners employed in the territory. This can be seen from the crews sent over to reinforce the garrison in the reign of Henry VII, which on five occasions solely consisted of gunners (see Table 12). By contrast, the crews of the second half of the reign of Edward IV had comprised larger numbers of men-at-arms and archers. As Grummitt has argued, this is evidence of greater specialisation in the personnel sent to the territory from 1487 to 1492.[124]

An examination of the roles and tasks undertaken by the gunners at Calais also shows that change occurred throughout the period. In the late fourteenth and early fifteenth centuries, the main gunners are consistently recorded as constructing artillery for the garrison. This can be seen with William Bretteville who supplied thirty-four guns between 1390 and 1404, and his successor, William Gerardson, who provided a further ten guns between 1404 and 1415.[125] After 1417, however, there is limited evidence for the construction of guns in the Pale itself. Instead guns were mostly purchased from England in the reign of Henry VI and from Flanders under Edward IV.[126] The main role of the later gunners appears to have been the production of gunpowder, which is recorded in the accounts of the treasurer on a regular basis.[127] They were also involved in the testing and firing of guns, such as in 1467–8, and in supervising labourers tasked with moving guns around the fortifications of Calais, as in 1447–50.[128]

121 Grummitt, *The Calais Garrison*, p. 106.

122 TNA, E 101/197/20, f. 2v; E 405/59, m. 8v; E 405/60, m. 4.

123 TNA, E 364/109, rot. D; E 364/112, rot. A dorse; E 101/55/23, ff. 5r, 8v.

124 Grummitt, *The Calais Garrison*, p. 59.

125 TNA, E 364/25, rot. E; E 364/28, rot. B; E 364/30, rot. E dorse; E 364/36, rot. D dorse; E 364/38, rot. D dorse; E 364/43, rot. H; E 364/44, rot. B; E 364/65, rot. A dorse.

126 An exception to this is the construction of thirty-four guns by Herman Donker in 1450–1, TNA, E 364/87, rot. H; E 403/725, m. 7; E 101/194/6, f. 27r; E 364/103, rot. E.

127 For example, TNA, E 364/72, rot. D dorse; E 364/87, rot. H; E 364/96, rot. B; E 364/106, rot. I.

128 TNA, E 364/103, rot. E; E 101/194/8, f. 37r.

TABLE 12 Other named gunners at Calais.

Name	Period of service	Number of years
Richard Selby	1384–7	3
Dederico van Ludyk	1384–7	3
William Sim'	1405–7	2
John Barton	1405–7	2
Hans Gunner	1415	1
John Bedford	1415–17	2
Richard Foriet	1415–17	2
John Boston	1434–6	2
John Terry	1436–9	3
Robert Lawney	1450–1	1
John Gunner	1456–7	1
Walter Humble	1466	1
John Dean	1466	1
Henry Frost	1482	1

Source: Information derived from TNA, E 364, E 159, E 101, E 403.

Another important role carried out by gunners was the operation of guns in wartime. It would appear, based upon the small number of gunners to guns, that they oversaw groups of labourers or soldiers who carried out the actual firing of these weapons. This is suggested by the entries in the accounts of the treasurer, which occasionally record that gunpowder and other equipment was expended by soldiers and gunners in defence of Calais, as in 1456-7. Gunners also took part in offensive operations carried out by the garrison, as in the attack against the Burgundian held castle of Alabon in 1471.[129] It would appear, based upon these sources, that gunners were employed more frequently in the use of artillery in the second half of the century. For instance only soldiers are recorded as having used guns in the offensive and defensive actions of 1406 and 1412.[130] By contrast, gunners and soldiers are more regularly described as employing guns in the defence of Calais in the reigns of Henry VI and Edward IV, as well as occasionally in sieges, such as at Hammes Castle in 1461.[131] This indicates that changes in the types

129 TNA, E 364/92, rot. C dorse; E 101/194/6, ff. 8r, 19r; Rainey, 'The Defence of Calais, 1436–1477', p. 98.

130 TNA, E 364/44, rot. B; E 364/65, rot. A dorse.

131 TNA, E 364/72, rot. D dorse; E 364/96, rot. C dorse; E 364/99, rot. C dorse; E 364/101, rot. H; E 364/107, rot. J; E 364/112, rot. C; E 101/195/14, f. 10r.

of guns used in the late fifteenth century meant that they were more difficult to operate, or alternatively that greater efforts were made to use them in an effective fashion. This development coincides with the use of new types of guns, notably the serpentine, which suggests that the two things were linked.

TABLE 13 Total recorded number of gunners at Calais.

Year	Number of gunners	Year	Number of gunners
1355–7	1	1456–7	2
1358–74	Not known	1458–65	1
1375–7	1	1466	3
1378–83	Not known	1467–8	2
1384–7	5	1469–72	1
1387–90	2	1472–3	2
1390–1404	1	1474–81	1
1405–7	3	1482	2
1407–15	1	1483–6	1
1415–17	4	1487	25
1417–34	Not known	1488	25
1434–6	3	1489	31
1436–9	3	1490	24
1440–9	1	1491–2	25
1450–1	3	1493–1500	1
1452–5	1	1501–2	10

Source: Information derived from TNA, E 364, E 101, E 403, E 404 (note that these numbers almost certainly underestimate the actual totals).

There is also some evidence that the status of some of the gunners at Calais changed over the course of the fifteenth century. Master gunners at Calais consistently received 12d. per day from the late fourteenth century to the end of the fifteenth century. In the 1380s, Thomas Gymour received 12d. per day, whereas his assistant received 3d.[132] Gymour's fifteenth-century successors, such as William Gerardson, 1410–11, and Henry Frost, 1482, continued to be in receipt of 12d. per day.[133] By

132 TNA, E 364/22, rot. H.
133 TNA, E 101/185/1, f. 59r; E 405/70, m. 1d.

contrast, the wages of the temporary gunners varied over time. In 1386, the gunners employed at Hammes Castle received 8d. per day.[134] Later, in December 1435, the two gunners sent to reinforce Calais, Herman Donker and John Boston, were each given a daily wage of 8d., with their two assistant gunners each getting 6d.[135] This had changed by the reign of Henry VII, with all of the gunners sent as part of crews to the Pale, including master gunners, only receiving 6d. per day.[136] Therefore, although the permanent master gunners retained their wages and status, the temporary gunners, perhaps because they were more numerous than before, became less valued.

This discussion has revealed that it is possible to detect broad changes in the activities and status of gunners in the Pale of Calais over the course of the fourteenth and fifteenth centuries. The main developments included longer periods of service on average by the main gunners at Calais, a move towards the purchase rather than construction of guns at the territory, together with a rise in the number of gunners, who tended to be from the Low Countries. Gunners also played an increased role in overseeing the operation of guns by teams of labourers or soldiers, which corresponds with changes in the types of guns used in the late fifteenth century. The quality of the surviving evidence means that a complete picture of the evolution of the gunner in Calais cannot be obtained, yet the available information demonstrates that important changes took place.

CONCLUSION

The artillery of the Calais garrison changed significantly between 1377 and 1488. The accounts of the treasurer reveal that the number of guns possessed by the garrison rose greatly during this time. It was not a steady rise, however, as large quantities of additional guns were acquired as a response to external threats, notably in 1385–6, 1436 and 1450–1. From 1471 onwards, Calais became the location of a large arsenal, which was used to furnish artillery for the major expeditions against France and Scotland. It was for this reason, as we have already seen, that the accounts of the treasurer become less reliable as a guide to the total number of guns stored in the territory, because a significant proportion of the artillery constructed in the Pale was administered separately by the victualler of Calais, William Rosse. The views and inventories also reveal that large numbers of guns were moved about between 1474 and 1488. This demonstrates the mobile nature of a significant proportion of the artillery stored at Calais, much of which was carted.

Important technological developments also occurred, particularly in the second half of the fifteenth century. The accounts of the treasurer, in general, provide limited

134 TNA, E 403/512, m.4.

135 TNA, E 403/721, m. 9.

136 TNA, E 404/79, no. 32; E 404/80, nos. 76, 259.

FIGURE 20 Proportion of iron to bronze guns at Calais 1375–1485.

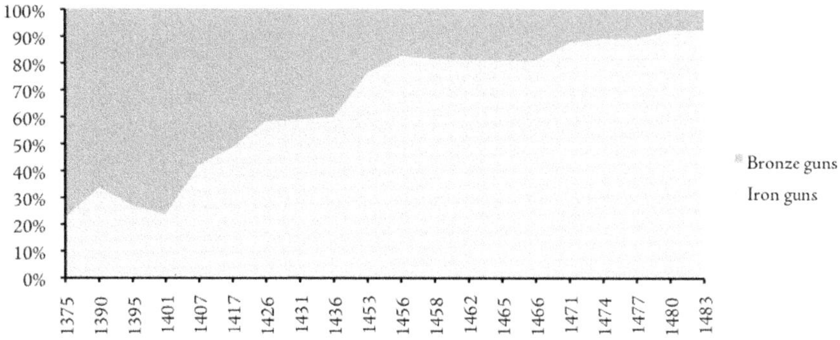

Source: Information derived from the enrolled accounts of the treasurer, TNA, E 364.

information as to the types of guns possessed by the garrison. Yet they do reveal that a notable shift occurred from the majority of guns being of cast bronze construction in the late fourteenth century to the majority being made of wrought iron by the end of the fifteenth century (see Figure 20). It is unclear if this was motivated by the cost advantage of using iron, which was cheaper than bronze, or whether the composite nature of iron guns meant that they were easier to repair. Furthermore, from the mid-fifteenth century onwards, an increasing variety of different guns are regularly mentioned in the accounts, including culverins, bombards, fowlers and organ guns.[137] The most significant new type of gun to emerge was the serpentine, which was first recorded in 1450. The long barrel of this weapon meant that it had greater range and accuracy, making it more effective in targeting enemy personnel. The views make clear that serpentines were an important weapon both offensively and defensively in the second half of the reign of Edward IV. Another notable development in the 1470s was the adoption of handguns. The earliest mention of handguns in Calais occurs in the year 1469, and they existed in large numbers in the 1470s and 1480s. These weapons were relatively light and easy to transport which meant that they were useful on expeditions, but could also be kept in storage in the fortifications of Calais ready for the use of the garrison. Other types of guns also came into use in this period, including pot guns, cart guns and bombardelles.[138] New techniques were necessary to ensure that these weapons were used efficiently, as can

137 It should be noted, however, that the accounts of the treasurer are slow to record the emergence of the fowler gun type, which is first mentioned in 1436 in Calais. By contrast, fowlers are first recorded in Exchequer documents in 1415, TNA, E 159/194, brevia directa baronibus, Michaelmas, rot. 6.

138 See the Glossary and Appendix J for a discussion of these terms.

be seen by the emergence of different types of gunpowder designed specifically for serpentines, hackbuts, bombards, as well as touch powder.[139] The Calais accounts, therefore, show that comparatively rapid changes in the types and ways that guns were used occurred in the second half of the fifteenth century.

The development of artillery fortifications show that these were intrinsically linked to the acquisition of guns. This can be seen in the case of Guînes Castle, a position whose defences were extensively remodelled between 1462 and 1474 and which was equipped with large numbers of guns. The diversification of gun types also meant that certain weapons became more suitable for deployment in particular locations, such as serpentines in elevated positions and fowlers on murderers. The Calais garrison attempted to improve the defences of the territory through adapting guns and fortifications in tandem. The example of Hammes Castle shows that the presence of new defensive features, such as bulwarks, was not always considered necessary for the effective defensive use of artillery. Nevertheless, the growing sophistication of artillery fortifications and guns in the second half of the fifteenth century demonstrates that important changes had taken place.

The position of the gunner in Calais was also transformed in the period. There was a growing tendency for the main gunners at the territory to serve for longer periods of time and to be from the Low Countries. This can be demonstrated by the lengthy careers of two men who appear to have been related to each other: Herman Donker, twenty-nine years, and William Donker, twenty-seven years. The role of the gunners also changed from focusing on the construction of guns to their administration and supervision. This indicates that the different gun types used in the second half of the fifteenth century required greater attention. Overall, the Calais evidence shows that the number and type of guns changed significantly in the second half of the fifteenth century. These developments resulted from royal interest in ordnance, changes in the use of artillery in warfare, and technological advances from Continental Europe. The changes meant that the gunpowder artillery of the Pale of Calais was transformed in the reign of Edward IV.

139 Serpentine powder is first recorded in the accounts in 1461–2, hackbut and bombard powder in 1475 and touch powder in 1481, TNA, E 364/72, rot. E; E 364/96, rot. C; E 101/198/13, f. 6r; E 364/119/36, rots. B, C.

5

Royal Castles and Guns

CASTLES WERE IMPORTANT STRUCTURES in late medieval England, serving a variety of functions, such as residences, prisons and administrative centres. They had also played a significant role in English warfare ever since their introduction by the Normans in the eleventh century. As fortified buildings, they could be used by defenders to protect a specific location from attack, or alternatively they could play an offensive role as a secure base from which a garrison could exert control over the surrounding area.[1] The security provided by castles also meant that they could be employed as arsenals and as supply depots for field armies.[2] This can be seen with the castles constructed at great expense by Edward I in the late thirteenth century to secure his conquest of North Wales.[3] These castles are large and impressive masonry structures, often with multiple circuits of walls, strategically sited near the coast or rivers. Later English success in crushing the rebellions led by Madog ap Llywelyn in 1294–5 and Owain Glyndŵr in the early fifteenth century was due, at least in part, to the possession of these castles, which could be easily supplied by sea, defended by small garrisons and used as forward bases to take the war to the Welsh rebels.[4] However, it was the final major programme of castle construction initiated by an English monarch. This was due, at least in part, to the significant increase in the number of castles in the possession of the crown in the late Middle Ages, notably those acquired from the estates of the duchies of Lancaster and York by the accessions to the throne of Henry IV and Edward IV in the fifteenth century.

The last royal castle constructed in the Middle Ages was Queenborough, situated on the Isle of Sheppey in Kent. This was built by order of Edward III

1 R. A. Brown, *English Castles* (Woodbridge: Boydell Press, 2004), p. 123. It is important to note that these castles were not built purely for military reasons, with cultural, symbolic, economic and administrative factors also being significant.

2 N. J. G. Pounds, *The Medieval Castle in England and Wales: A Social and Political History* (Cambridge: Cambridge University Press, 1990), p. 125.

3 Brown, *English Castles*, p. 64.

4 R. R. Davies, *The Revolt of Owain Glyn Dŵr* (Oxford: Oxford University Press, 1995), p. 249.

between 1366 and 1377, with its concentric design strongly indicating that it was designed to provide coastal defence, further evidence of which can be seen by the provision of artillery for its safeguard.[5] By the late Middle Ages, however, most English castles were very rarely, if ever, involved in military operations, except for those in frontier regions such as near the Scottish border. As a result of this, the vast majority of castles, even important royal ones, were only garrisoned during times of crisis, notably in 1385–6.[6] During times of peace, the establishments of these buildings were generally very small; for instance Robert Bardolf was appointed constable of Portchester Castle in 1381, receiving wages just for himself, a porter, a groom, an artillier and a watchman.[7] Yet stores of arms and armour were kept in some royal castles, as can be seen from surviving inventories, with an account for 1372 to 1375 listing various items at Dover Castle, including 100 haubergeons, 100 kettle helmets, 100 jackets, 100 lances, 100 bows, and 12 crossbows.[8] These stocks were augmented by deliveries from the privy wardrobe when the safety of these fortifications appeared to be threatened, with Portchester Castle receiving 2 springalds, 50 bows, 10 axes and 20 lances in 1379, with 12 crossbows and 1,000 quarrels sent later, in 1386.[9] Therefore these structures still had a military function in this period, which explains why much of the earliest evidence for English guns concerns the provisioning of royal castles.

The weaponry stored at royal castles can be traced through multiple sources, with the most useful type of document consisting of the enrolled or particulars of account submitted by the constables of these fortifications to the Exchequer. Most of these castle inventories cease to be recorded by the Exchequer after the fourteenth century, with the main exception being Dover Castle, which was last inventoried in 1437.[10] Further information can be extracted from accounts of works, such as those drawn up by John Skipton, clerk of the works for the king, for ordnance provided to the castles and towns of Berwick, Carlisle and Roxburgh in the years 1426–34.[11] The records of the privy wardrobe also list the delivery of artillery to royal castles, yet the last known set of accounts for this office dates from the early fifteenth century. Therefore, most of the surviving evidence exists for guns in the late fourteenth century, which makes studying the fifteenth century difficult. It is possible, however, to glean limited information from occasional references in

5 Brown, *English Castles*, p. 94.

6 For example, see TNA, E 403/512, m. 19.

7 *CPR 1377–1381*, p. 594.

8 TNA, E 364/16, rot. C dorse.

9 Thom Richardson, *The Tower Armoury in the Fourteenth Century* (Leeds: Royal Armouries, 2016), pp. 127, 139, 156–7, 172.

10 TNA, E 364/73, rot. H.

11 TNA, E 364/74, rots. N–N dorse.

other Exchequer records, such as the issue rolls, warrants for issue and the tellers' rolls, as well as the accounts of the chamberlains of north and south Wales.

EDWARD III TO RICHARD II

The earliest known instance of guns being provided for the furnishing of an English castle occurs in 1365, when 2 large guns and 9 smaller guns were sent to Queenborough Castle from the privy wardrobe.[12] It is unlikely that this delivery was intended solely to strengthen the defences of the fortification, however, as the Treaty of Brétigny was still in force. Instead it seems probable that Edward III intended to show off his new castle by equipping it with the latest weaponry available; which is suggested by his visit to Queenborough in 1373, when its guns were fired in his presence.[13] The resumption of war in 1369 and a series of military defeats in France, however, gradually led to a small number of other castles in southern England being furnished with gunpowder artillery to improve their defences.[14] An early example of this is Dover Castle, which was provided with 6 guns in 1371.[15] No further firearms were provided by the privy wardrobe until 1379, however, when 3 were sent to Portchester Castle. The following year saw the delivery of 4 guns to Corfe Castle, 4 to various castles in Cornwall and 3 to Dover Castle together with large quantities of lead shot, in addition to 12lbs of gunpowder given to Southampton Castle and 100lbs of saltpetre to Carisbrooke Castle.[16]

During the same period, works were also carried out at most of these castles, which, at least in part, were intended to adapt their defences for the deployment of guns. At Southampton the castle was entirely rebuilt at a cost of almost £2,000 between 1378 and 1388, with £300 being spent on repairs at Corfe Castle at around the same time.[17] The lack of physical evidence means that it is very difficult to assess whether these castles were built with gunports, although the delivery of guns to both of the fortifications would suggest so. By contrast, it is still possible to see how Portchester Castle was adapted for the use of gunpowder weapons at this

12 T. F. Tout, 'Firearms in England in the Fourteenth Century', *The English Historical Review*, 26 (1911), 666–702 (p. 692).

13 Brown, *English Castles*, p. 95.

14 Dan Spencer, 'Royal Castles and Coastal Defence in the Late Fourteenth Century', *Nottingham Medieval Studies*, 61 (2017), 147–70.

15 Tout, 'Firearms in England in the Fourteenth Century', p. 692.

16 Ibid., pp. 695, 699, TNA, E 364/16, rot. C dorse.

17 H. M. Colvin, *The History of the King's Works*, volume 2 (London: H. M. S. O., 1963), pp. 623, 842–4; TNA, E 403/462, m. 15; E 403/465, mm. 6, 8, 17, 19; E 403/468, mm. 8, 17, 18; E 403/472, mm. 16, 17; E 403/475, mm. 21, 22; E 403/478, mm. 2, 19.

time. Between 1377 and 1381 a new tower, called Assheton's Tower, was constructed in the inner ward of the castle, with a number of keyhole insertions carved into the walls for the deployment of guns. The gatehouse of Carisbrooke Castle was also raised in height at this time, and included gunports that are still extant.[18] These developments were a response to the worsening security of the south coast, with French raids on the Isle of Wight and Rye in 1377, as well as on Winchelsea in 1380.[19] Ordnance was also sent to reinforce English-held castles in Brittany at this time, with 10 guns being purchased for the town and castle of Cherbourg, in addition to gunstones and saltpetre sent to Brest.[20]

The biggest increase in the number of guns sent to these fortifications occurred in the mid-1380s as a response to the threat of invasion by the French and their Scottish allies.[21] In January 1384 the Scots captured the English-held castle of Lochmaben before launching a raid into Cumberland.[22] Almost certainly as a response to this attack, the decision was made to improve the security of the border region by the government of Richard II. By a writ of 15 February of the same year, 4 guns with 60lbs of saltpetre were delivered to Roxburgh Castle, in addition to 8 guns, 60lbs of gunpowder and 40lbs of saltpetre to the town and castle of Berwick. Subsequently, on 20 March, a commission was appointed to assess the artillery and fortifications of Berwick and Roxburgh Castle, as a result of which one further gun was purchased for the defence of Roxburgh ten days later.[23] In the summer of 1385, Richard II led a large army in an invasion of Scotland, but this failed to deter Franco-Scottish plans for an invasion of England. A subsequent counter-attack by the allies resulted in the capture of Wark Castle in Northumberland, but an attack on Berwick was beaten off by the garrison, who expended 1 gun, 360 gunstones and 143lbs of gunpowder in this action.[24] The threat of attack meant that significant quantities of ordnance were purchased for the defence of the castles in the border region in 1385–6. This consisted of 8 guns, including 3 handguns, for the defence

18 John R. Kenyon, 'Early Artillery Fortifications in England and Wales: A Preliminary Survey and Reappraisal', *Archaeological Journal*, 138 (1981), 205–40 (p. 210).

19 Jonathan Sumption, *The Hundred Years War: Volume III, Divided Houses* (London: Faber, 2009), p. 281; James Sherborne, 'The Defence of the Realm and the Impeachment of Michael de la Pole in 1386', in *Politics and Crisis in Fourteenth-Century England*, ed. John Taylor and Wendy Childs (Gloucester: Alan Sutton, 1990), pp. 99–118 (p. 97).

20 TNA, E 364/18, rot. E; E 403/475, m. 20.

21 Jonathan Sumption, *The Hundred Years War: Volume III, Divided Houses* (London: Faber, 2009), p. 281; James Sherborne, 'The Defence of the Realm and the Impeachment of Michael de la Pole in 1386', in *Politics and Crisis in Fourteenth-Century England*, ed. John Taylor and Wendy Childs (Gloucester: Alan Sutton, 1990), pp. 99–118 (p. 97).

22 Jonathan Sumption, *The Hundred Years War: Volume III, Divided Houses* (London: Faber, 2009), p. 518.

23 *CPR 1381–1385*, pp. 424–5; TNA, E 364/27, rots. F–F dorse.

24 Sumption, *The Hundred Years War: Volume III*, pp. 547–8.

of Berwick, in addition to 3 guns for Carlisle and a quantity of gunpowder for Roxburgh.[25]

French plans for an invasion of southern England in the years 1385 and 1386 also meant that royal castles in this area were sent sizeable quantities of guns. Ordnance was acquired for Carisbrooke Castle with the wages of 6 gunners, who owned 8 guns, paid in November 1384 for the safeguard of the Isle of Wight.[26] As many as 12 guns were purchased in 1385 for the defence of Dover Castle by its constable, Sir Simon Burley, with another 8 guns provided from the privy wardrobe in 1386. Later the same year, a further 3 guns were sent to Portchester Castle and a further 2 to Southampton Castle.[27] At the latter place a chaplain, Thomas Tredyngton, was appointed in July 1386 to oversee the castle's armoury 'expressively because he is an expert in guns and the management of artillery'.[28] The threat of invasion eventually passed due to French delays in launching their expedition and English naval victories in 1387. Thereafter a reduction in hostilities, particularly after the signing of the Truce of Leulinghem in 1393, meant that only limited quantities of equipment were sent to royal castles for the remainder of the century. The decision to send three guns to Roxburgh Castle in 1397 appears to have been prompted by continuing tensions with Scotland.[29] In the late fourteenth century, therefore, firearms came to play an important part in the defence of royal castles in England, with attempts being made to adapt these fortifications for guns through the creation of gunports.

HENRY IV TO HENRY V

The upheaval caused by the usurpation of Henry IV in 1399 contributed to the outbreak of major rebellions in Wales and the north of England, with castles playing an important role in these conflicts. This was particularly the case in Wales where a long legacy of conflict with the native Welsh meant that the region was heavily populated with castles. A significant number of these castles were garrisoned by the English during the rebellion, and, although many of them were in a poor state of repair, they were a key aspect of royal military strategy.[30] Those inventories which survive demonstrate that at least some castles in Wales possessed these weapons

25 TNA, E 364/26, rots. A–A dorse; Tout, 'Firearms in England in the Fourteenth Century', p. 699; E 101/40/6; E 364/20, rot. A.

26 TNA, E 403/505, m. 9.

27 Tout, 'Firearms in England in the Fourteenth Century', pp. 698–9.

28 *CPR 1385–1389*, p. 196.

29 Sumption, *The Hundred Years War: Volume III*, pp. 676–8, 844–5; TNA, E 364/23, rot. B; E 364/32, rot. E dorse; Tout, 'Firearms in England in the Fourteenth Century', pp. 700–1.

30 Davies, *The Revolt of Owain Glyn Dŵr*, pp. 248–51.

by the time of the uprising. Surveys taken of the castles of Oswestry and Holt after their forfeiture to the crown in 1397 reveal that they collectively had three guns in their stockpile of arms, with a further four guns and other equipment sent to Holt Castle from the privy wardrobe in the following year.[31] Modest efforts were made to reinforce royal fortifications in north Wales in the early years of the rebellion, with very small quantities of gunpowder being sent to the castles of Chester, Flint and Rhuddlan between 1400 and 1403.[32] A master gunner, John Rouse, and carpenters were also employed on the repair of guns within castles in north Wales during this time.[33]

Rebel successes, such as the capture of the castles of Harlech and Aberystwyth, later led to more substantial deliveries of arms, with two guns and 40lbs of gunpowder being sent to Kidwelly Castle and a further six to Brecon Castle in 1404.[34] It is likely that these weapons played a part in allowing their garrisons to resist rebel attacks, thereby contributing to the eventual English success in suppressing the rebellion. Guns were subsequently left in the stores of the major castles in Wales, with one gun called *Clyff* and 950lbs of gunpowder left at Aberystwyth, two guns at Brecon, five guns weighing 200lbs and 200lbs of gunpowder at Pembroke, 700lbs of gunpowder at Cardigan, 950lbs of gunpowder at Criccieth, two guns at Carmarthen and two named guns at Harlech, *God's Grace* and *Fowler*, together with 5,700lbs of gunpowder.[35] By contrast, little effort was made to furnish royal castles in southern England with guns during the reign of Henry IV. The main exception to this was Sandwich Castle, which was sent two guns and 100lbs of gunpowder from the privy wardrobe in 1403, but this equipment was returned eight years later.[36]

Henry IV was also forced to contend with risings in the north of England led by the powerful Percy family, with Berwick captured and burnt by the Scots in 1405. The subsequent royal siege of the town and castle also caused further damage, as is revealed by letters sent by John of Lancaster, the third son of Henry IV, to the king's council. In a letter of 1406 he stressed the poor state of the fortifications of Berwick and the need to restock the castle with guns. By 1414 this had still to be addressed, when he once again wrote to the council, emphasising that 'The walls of the town and castle of Berwick were in a ruinous condition and were now fallen in

31 H. C. M. Lyte, ed., *Calendar of Inquisitions Miscellaneous*, volume 6 (H. M. S. O., 1916), pp. 108, 112–13; Tout, 'Firearms in England in the Fourteenth Century', p. 701.

32 TNA, SC 6/774/14, m. 6d; SC 6/775/3, m. 2d; SC 6/774/15, m. 1d; SC 6/774/15, m. 1d; SC 6/774/14, m. 4d.

33 TNA, SC 6/1216/1.

34 TNA, DL 28/4/3; James Hamilton Wylie, *History of England Under Henry the Fourth, Vol. IV. 1411–1413* (London: Longmans, Green and Co., 1898), pp. 187–8.

35 TNA, E 364/49, rots. C–C dorse.

36 TNA, E 364/43, rot. A; E 364/46, rot. E.

many places … and that the cannon, gunpowder, arms, artillery, and victuals in the town had been insufficient ever since the fire and the rebellion of Northumberland'.[37] Unfortunately no inventories survive for Berwick for this period, but another account records that 250lbs of gunpowder was placed in the stores of the castle after the siege, in addition to a further 250lbs of gunpowder sent to Warkworth Castle.[38] Roxburgh also appears to have been affected by the rebellion and Scottish attacks, with an inventory of 1410–11 revealing that the number of guns at the castle had fallen slightly from five to four.[39] Five years later a commission was given in 1416 to Sir John Clavering and Robert Harbottell to survey the state of the fortifications together with the artillery at the castle.[40] This document listed serious defects in the defences of the castle, yet the garrison were able to resist a Scottish siege the following year.[41] Nevertheless the need was felt to reinforce the weaponry of the fortress, with £26 13s. 4d. being paid for the purchase and carriage of guns to Roxburgh on 15 November, and a further £20 expended for the same purpose in July of the following year.[42] Berwick was also reinforced two years later, with £26 15s. 2d. being spent on purchases of guns, gunpowder and other ordnance.[43] Guns therefore continued to play an important part in the defence of royal castles in the fifteenth century. No evidence survives, however, for works being carried out to adapt these fortifications for the use of guns, although it seems highly likely that gunports and other minor changes were made to castles such as Roxburgh and Berwick.

HENRY VI TO HENRY VII

The possession of Normandy and domestic peace meant that there was little need for guns to be sent to royal castles in England and Wales for most of the first half of the fifteenth century. Surviving inventories indicate that these fortifications still possessed old guns but that no new weaponry was acquired; with Dover Castle possessing eighteen guns between 1403 and 1437 and Southampton Castle two guns between 1437 and 1442.[44] Stockpiles of artillery were also kept in Wales, with nineteen guns listed in the accounts for the chamberlain of north Wales for 1425–30

37 S. B. Chrimes, 'Some Letters of John of Lancaster as Warden of the East Marches Towards Scotland', *Speculum*, 14 (1939), 3–27 (pp. 7, 12).

38 TNA, E 364/49, rot. C dorse.

39 TNA, E 364/23, rot. E dorse; E 364/46, rot. D dorse.

40 TNA, C 47/2/49, no. 23.

41 Colvin, *The History of the King's Works*, volume 2, pp. 820–1.

42 TNA, E 403/633, m. 4; E 403/636, m. 9.

43 TNA, E 403/636, m. 1.

44 TNA, E 364/37, rot. D; E 364/50, rot. A; E 364/65, rot. H dorse; E 364/73, rot. H; E 364/80, rot. H; E 364/76, rot. M.

and two pellet guns in those of the chamberlain of south Wales in 1435–7.[45] Later in 1449, the threat of invasion prompted the inhabitants of the Isle of Wight to send a petition to the crown requesting military equipment.[46] In response, a substantial quantity of ordnance was sent to Carisbrooke Castle, consisting of eight fowlers, six serpentines, twelve cart guns and twenty-four culverins at a cost of £60.[47] The warrant for issue states that this weaponry was sent due to the 'perille that oure ysle of Weight standeth in … also the unsuffisaunce of stuff of ordenuaunce, artillarye and habiliments of werre withunce oure castel of Caresbroke'. A writ of 9 May 1452 records that a John Slyfirst, yeoman of the artillery within the castle, hired two carters to carry this equipment from London to Southampton, where it was shipped to Newport and then on to Carisbrooke Castle.[48]

Later in the decade, Kenilworth Castle was used as an arsenal for the field guns of the Lancastrian government, with a Stephen Clampard being paid £5 for his expenses in having transported guns to the castle on 21 October 1456. In the following January, John Judde, master of the king's ordnance, was paid for bringing into 'our castell of Kenelleworth xxvj newe gunnes called serpentynes for the felde' together with other equipment. Kenilworth lacked the facilities to maintain these weapons, however, with four serpentines and twenty-one cart guns sent from there to the Tower of London to be stocked and carted in September 1458.[49] The chronicler John Benet records that at a meeting of the king's council held in Coventry in 1460, it was decided that the fortifications of Kenilworth should be improved. Instructions were also issued at the same time for guns to be sent there from the Tower of London.[50] Benet's testimony regarding the fortifying of the castle is seemingly supported by a later set of accounts for the lordship of Kenilworth for 1483–4, which mentions that repairs were carried out to the bulwark and a 'gun tower' at that time.[51] Limited efforts were also made to furnish other castles with

45 TNA, E 364/64, rot. E; E 364/73, rot. N.

46 TNA, SC 8/28/1353.

47 Dan Spencer, 'The Lancastrian Armament Programme of the 1450s and the Development of Field Guns', *The Ricardian*, 25 (2015), 61–70 (p. 62).

48 TNA, E 404/66, no. 216; E 159/229, brevia directa baronibus, Trinity, rot. 14d; E 403/783, m. 2.

49 Spencer, 'The Lancastrian Armament Programme of the 1450s and the Development of Field Guns', pp. 65–6; TNA, E 404/71/3, no. 1.

50 Harriss, G. L., and M.A. Harriss, eds., 'John Benet's Chronicle for the years 1400 to 1462', in *Camden Miscellany XXIV*, Camden 4th ser., vol. 9 (1972), p. 225.

51 TNA, DL 29/463/7564. It is difficult to determine when exactly the bulwark and gun tower were constructed at Kenilworth, as the accounts for the lordship rarely mention the names of specific buildings when recording the amounts spent on carrying out repairs to the castle. The roll for 1483–4 is comparatively unusual in that it records the names of multiple structures that were repaired during this time. A date of c.1456–60 is supported by Benet's testimony and the Exchequer evidence for the transfer of guns

ordnance in the late 1450s, with a commission granted to three men in September 1457 to procure guns and gunpowder for Dover Castle, in addition to a delivery of gunpowder to Bristol Castle at around this time.[52] In the following year William Waynflete, bishop of Winchester, wrote to Judde to request that 'stuffe to be hadde in to the castell of Hampton for keping therof', which suggests that its arsenal of weaponry was considered to be inadequate.[53] Judde was later ordered on 1 December 1459 to survey the condition of all the ordnance in castles and towns in England, following the flight of the Yorkist lords after the battle of Ludford Bridge, to ensure that they were ready for defence.[54] Small quantities of firearms were also kept in castles in Wales in the 1450s and 1460s, with two handguns and 16lbs of gunpowder being purchased for Rhuddlan Castle in 1465–7.[55]

Few accounts of work survive for the late fifteenth century, but occasional references to artillery fortifications in other financial documentation indicate that efforts were made to modify the defences of at least some royal castles in the reign of Edward IV.[56] A letter given under the privy seal, which was later enrolled in the King's Remembrancer's Memoranda Roll, dated 1 August 1476, states that the bailiffs of Sandwich should be remunerated for having made payments by royal command from Michaelmas 1472 to Michaelmas 1474.[57] These expenses include works that were carried out at Sandwich Castle, which principally consisted of the 'castyng of the castell diche', as well as the construction of a drawbridge. Interestingly, reference is also made to payments for the 'settyng of lowpes within the castell diche for gunnes', as well as the cost of purchasing and transporting timber for their construction. The mention of timber loops within the 'diche' suggests that this was not a ditch but instead a small mound of earth known as a 'falsebraye' erected in front of the wall of the castle. These defences were intended to provide additional protection for brick or stone walls by absorbing gunshot, as well as allowing the defenders to fire from its loops. Early in the following decade, a brick bulwark was

to the castle. Yet it is also possible that they date from the reign of Edward IV and could be contemporary with the construction of the bulwark outside the Tower of London. For a discussion of the latter see below.

52 Spencer, 'The Lancastrian Armament Programme of the 1450s and the Development of Field Guns', pp. 65–6; E. W. W. Veale, ed., *The Great Red Book of Bristol*, volume 1 (Bristol: Bristol Record Society, 1931–53), p. 131.

53 R. C. Anderson, ed., *Letters from the Fifteenth and Sixteenth Centuries: From the Archives of Southampton* (Southampton: Cox & Sharland, 1921), pp. 12–13.

54 Spencer, 'The Lancastrian Armament Programme of the 1450s and the Development of Field Guns', pp. 65–6.

55 TNA, SC 6/778/12, m. 3d; SC 6/779/10, mm.1d, 6d; SC 6/780/5; SC 6/780/8, m. 4d.

56 For example, one of the few places for which accounts of building work survive for the reign of Edward IV is Nottingham Castle, for which see TNA, E 101/478/15; E 101/478/16.

57 TNA, E 159/253, recorda, Michaelmas, rot. 2d.

also constructed to protect the entrance to the Tower of London.[58] Archaeological excavations reveal that the structure consisted of a hollow bastion with walls approximately one metre thick.[59] In 1484, Thomas Redhede was appointed as porter of the Tower of London and as the keeper of the bulwark outside the west gate of the Tower with a daily wage of 6d.[60] Eleven years later, an inventory records that three guns were stored in the bulwark.[61] There is also limited evidence that castles were used to keep military equipment, with a gunpowder store established at Southampton in 1492 for the expedition to Boulogne.[62]

By contrast, guns were supplied on a more regular basis to royal castles on the Anglo-Scottish borders. Relations with Scotland had gradually deteriorated after the release of James I from English captivity in 1424, leading to the outbreak of war in 1436.[63] This prompted the government to greatly reinforce the defences of the key northern fortifications, as can be seen from the account of John Skipton, clerk of the works for the king, for the years 1426–34.[64] This included the purchase of twelve guns for Berwick and Roxburgh together with six guns for Carlisle. An additional quantity of four guns was given to Berwick, six to Carlisle and five to Roxburgh, along with gunpowder, saltpetre and gunstones plus other weapons. Furthermore, in May 1434, £54 was provided by the Exchequer for the purchase of artillery and £54 for the repair of Roxburgh Castle.[65] The commencement of hostilities, in May 1436, resulted in the despatch of additional quantities of gunpowder.[66] These were required as Roxburgh was besieged by a Scottish army led by James I, which was well equipped with artillery.[67] Following the lifting of the siege, a further three barrels of gunpowder was sent to replenish the castle's stores, together with bows, arrows and other equipment, in addition to the construction of a bulwark at Berwick.[68]

A later account of works for the period 1446–50 records the refurbishment of guns in Berwick, with a gunner, John Long, being paid to repair a gun kept at Saint

58 TNA, E 101/517/23, no. 100.

59 Michael Hutchinson, 'Edward IV's Bulwark: Excavations at Tower Hill, London, 1985', *Transactions of the London and Middlesex Archaeological Society*, 47 (1996), 103–44 (pp. 108–9).

60 *CPR 1476–1485*, p. 380.

61 TNA, E 36/8, f. 90v.

62 TNA, E 36/285, f. 5.

63 Griffiths, *The Reign of King Henry VI*, pp. 155–61.

64 TNA E 364/74, rot. N.

65 TNA, E 404/50, no. 295.

66 TNA, E 403/723, m. 14; E 404/52, no. 387; E 364/75, rots. G–G dorse.

67 Ralph A. Griffiths, *The Reign of King Henry VI: The Exercise of Royal Authority, 1422–1461* (London: Benn, 1981), pp. 161–2.

68 TNA, E 404/53, no. 173; E 364/75, rots. G–G dorse.

Mary's Gate.[69] However, there is no further evidence for the provision of firearms for these castles for the next three decades, which, it has been argued, was due to responsibility for procuring ordnance being devolved to the wardens of the marches in the 1440s and 1450s, who only submitted summary accounts to the Exchequer.[70] Later in the 1480s, conflict with Scotland prompted the carrying out of measures to improve the defences of the region. On 18 May 1480, instructions were issued by the king for bombards and other guns to be transported from Nottingham Castle to Norham Castle.[71] The sum of £2 was subsequently spent on the wages of smiths, carpenters and other workers for making of all manner of 'necessaries as bolewarks frames and beds for gonnes and for tymbr and carriage nedefull for the sure keeping of the castell'.[72] Guns, gunpowder and other equipment was also sent to Berwick Castle in 1482 and 1483, with a survey of artillery stored in northern castles carried out in 1488.[73] In 1490 over £1,100 was spent on the purchase of ordnance for the towns and castles of Berwick, Carlisle and Scarborough.[74] Later in 1497, a substantial portion of the royal artillery sent for the expedition to Scotland was subsequently left behind in Berwick Castle after the end of the campaign. This totalled eighteen guns consisting of one curtow, two demi-curtows, two serpentines, two small serpentines and eleven falcons, in addition to ammunition, horse harnesses and gunpowder.[75] These weapons were overseen by gunners in royal pay, with a master of the ordnance employed at Berwick from 1487 to 1493 with a team of twenty gunners.[76]

69 TNA, E 101/483/11, m. 11.

70 David Grummitt, 'A Military Revolution in the North? The Impact of Gunpowder Weaponry on the Anglo-Scottish Marches', in *England and Scotland at War, c. 1296–c. 1513*, ed. Andy King and David Simpkin (Leiden-Boston: Brill Academic Publishers, 2012), pp. 283–96 (p. 290).

71 *CPR 1476–1485*, p. 213.

72 DUL, CCB B/23/2/2.

73 This included a delivery of twelve serpentines for the town and castle of Berwick in 1488. TNA, E 159/265, brevia directa baronibus, Michaelmas, rot. 24. *CPR 1476–1485*, p. 462; E 405/70, m. 7d; E 405/75, m. 14r; Joseph Bain, ed., *Calendar of Documents Relating to Scotland. Volume 4, A.D. 1357–1509. Addenda–1221–1435* (Edinburgh: H. M. General Register House, 1888), p. 315.

74 TNA, E 405/78, mm. 28d, 31d, 34r–35r.

75 M. Oppenheim, ed., *Naval Accounts and Inventories of the Reign of Henry VII, 1485–8 and 1495–7* (London: Naval Records Society, 1896), p. 127.

76 TNA, E 159/265, brevia directa baronibus, Pascha, rots. 1–2; Bain, *Calendar of Documents Relating to Scotland, Volume 4. A.D. 1357–1509*, p. 315.

CONCLUSION

Guns were used in the defence of English royal castles from the mid-fourteenth century into the early modern period, but their role changed significantly over time. In the reign of Richard II, guns were deployed to a number of key fortifications in southern and northern England to enhance the firepower of their garrisons, with their defences being adapted through the addition of gunports. This policy continued into the early fifteenth century, with guns regularly supplied to furnish the arsenals of threatened castles in Wales and on the Anglo-Scottish marches. Later in the 1450s, the danger of French raiding prompted the despatch of ordnance to furnish castles on the south coast. Yet there is little evidence for the presence of firearms in royal fortifications in England, except for those on the border with Scotland, in the second half of the fifteenth century. This may be explained, at least in part, by gaps in the surviving sources, particularly with accounts of works, few of which are extant for the reign of Edward IV. Isolated references to the construction of artillery fortifications at Kenilworth Castle, Sandwich Castle and the Tower of London in the 1460s, 1470s and 1480s, demonstrate that at least some castles were adapted for gunpowder weapons in the second half of the fifteenth century. These works mirror the development of specialist artillery fortifications in the Pale of Calais, albeit on a much smaller scale. Therefore, castles continued to be modified to meet changing needs throughout the fifteenth century and still retained their military value. Nevertheless, there is far more evidence for works carried out to improve urban defences, which indicates that the onus of coastal defence had, on the whole, shifted from royal castles to the towns, with English kings keen to promote this development. This change in royal policy may have been motivated by financial considerations, as well as by the increasing numbers of incorporated towns in the fifteenth century, who thereby assumed greater responsibilities for their own defence.

Part II

Urban

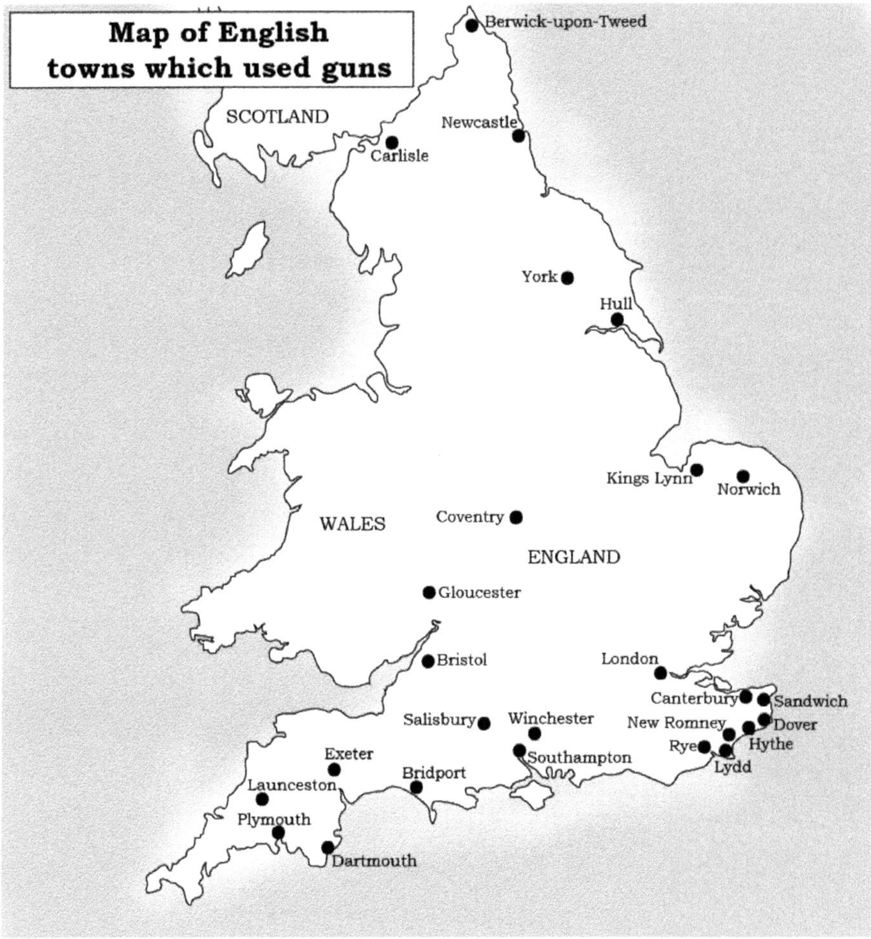

Map 2 English towns with evidence of gun usage 1360–1500.
Map drawn by Scott Hall © Dan Spencer.

6

Towns

IN THE LATE MIDDLE Ages, towns contained nearly a fifth of England's population and were important economically to the wealth of the realm.[1] These settlements varied considerably in size, from York, for instance, which had a population of 10,000 at the beginning of the fifteenth century, to much smaller places, such as Lydd in Kent, with only 1,000 inhabitants.[2] The economic development of these urban centres also differed markedly, with settlements such as York, which had traditionally been considered the second most important city in the country, in decline by the fifteenth century.[3] By contrast, towns such as Norwich and Exeter were growing in size and wealth during the same period.[4] English towns had long been in receipt of important privileges from the crown, which provided them with limited independence from royal interference in their internal affairs.[5] The fifteenth century saw an increase in the process of incorporation, whereby a select number of towns were able to successfully petition for important legal rights. These towns thereby acquired a legal status that allowed them a degree of self-governance, as can be seen with Southampton's charter of 1445, which stated that the settlement should henceforth have a mayor, two bailiffs and the burgesses.[6] The Cinque Ports, a confederation of ports in the south and east of England, had long enjoyed special privileges from the crown in return for supplying fifty-seven ships for fifteen days each year at their own expense.[7]

1 Lorraine Attreed, *The King's Towns: Identity and Survival in Late Medieval English Boroughs* (New York: Peter Lang Publishing, 2001), p. 1.

2 Ibid., p. 2; Spencer Dimmock, 'Class and the Social Transformation of a Late Medieval Small Town: Lydd c. 1450–1550' (unpublished doctoral thesis, University of Kent, 1998), p. 25.

3 Attreed, *The King's Towns*, p. 2.

4 Ibid., pp. 2–3.

5 Ibid., p. 35.

6 Martin Weinbaum, *The Incorporation of Boroughs* (Manchester: Manchester University Press, 1937), pp. 45, 63; Colin Platt, *Medieval Southampton: The Port and Trading Community, A.D. 1000–1600* (London: Routledge and Kegan Paul, 1973), pp. 165–7.

7 K. M. Elisabeth Murray, *The Constitutional History of the Cinque Ports* (Manchester: Manchester University Press, 1935), p. 24.

The rights acquired by the towns were balanced by responsibilities, notably the obligation to provide adequately for their own defence. This was an important task which could have serious consequences for civic authorities if they were thought by the king to be negligent, as occurred at Southampton in 1338 in the aftermath of a French raid, when Edward III temporarily rescinded the privileges of the town and placed it under the control of a royally appointed keeper.[8] Security often took the form of 'watch and ward', which was a system whereby towns were divided into wards whose inhabitants were responsible for providing manpower for service as watchmen, guards and patrols, as well as expenditure on military equipment, such as guns.[9] In addition a sizeable minority of English towns in the later Middle Ages possessed walls; for instance by 1520, 108 of the 249 towns with charters were walled.[10] Walls were important in providing protection to towns, as well as contributing to civic pride and prestige; nevertheless, a number of important settlements, such as Beverley and Salisbury, were never walled.[11] Towns were also called upon to make contributions to the military campaigns of English kings, with Salisbury, for example, providing fifty men-at-arms and twenty-four archers for the defence of Calais in 1451.[12] One of the most notable changes in the late Middle Ages was the adoption of gunpowder weapons by a sizeable number of towns. This chapter will examine all the English towns that have extant evidence for the use of guns in the period.

Military activity can be traced through the extant financial records for the towns, and, although the survival rate of these documents is rather uneven, sufficient evidence exists to trace the changes that occurred throughout the period. A variety of different financial records survive for the towns including chamberlains' accounts, stewards' accounts and minute books, which record income and expenditure by year. The process of financial record-keeping can be seen in the Kentish town of Lydd, which was a corporate member of the Cinque Ports, as a 'limb' of New Romney.[13] Financial accounts survive in book form for the years 1428 to 1484. It is possible that regular accounts were not kept prior to this date as the first line in the book reads 'Assit principio sancta maria meo'

8 Randall Moffett, 'The Military Organization of Southampton in the Late Medieval Period, 1300–1500' (unpublished doctoral thesis, University of Southampton, 2009), p. 168.

9 O. H. Creighton, *Medieval Town Walls: An Archaeology and Social History of Urban Defence* (Stroud: Tempus, 2005), p. 188.

10 Hilary L. Turner, *Town Defences in England and Wales: An Architectural and Documentary Study AD 900–1500* (London: John Baker Ltd, 1971), p. 91.

11 Ibid., p. 92; Creighton, *Medieval Towns Walls*, p. 207.

12 David R. Carr, ed., *The First General Entry Book of The City of Salisbury 1387–1452* (Trowbridge: Wiltshire Record Society, 2001), p. 233.

13 Murray, *The Constitutional History of the Cinque Ports*, p. 13.

(Assist my beginnings, Saint Mary). The start of the accounting year varied year by year until 1476 when it was fixed as ending on the feast of St Mary Magdelene on 22 July. Prior to 1477, twelve jurats managed the financial affairs of Lydd, with six jurats leaving office each year to be replaced by newly elected men; after this date responsibility for financial affairs was managed by two chamberlains who were elected annually.[14] The annual accounts record a great variety of items of expenditure for the town including expenses for horses, ships, food, gifts as well as defence.[15] Every year these accounts were audited at the Hundred Court in the presence of the bailiff and jurats.[16] These types of sources provide plentiful information on how towns used gunpowder weapons.

EARLY EVIDENCE FOR TOWNS

English towns had made use of artillery for their defence long before they acquired gunpowder weapons, with the town of Norwich possessing thirty springalds in 1342.[17] The deployment of guns from the privy wardrobe to royal castles in the second half of the fifteenth century, however, appears to have encouraged the towns to adopt this technology.[18] It is striking that many of the towns which had guns stationed in their royal castles between the 1380s to early 1400s, such as Dover, Rye, Sandwich and Southampton, went on to make considerable use of gunpowder weapons in the second half of the fifteenth century, as we shall see. Norwich was an early adopter of gunpowder weapons, with a muster of 1365 including two gunners with their equipment, although no evidence survives of royal guns in Norwich Castle.[19] The chamberlain accounts for 1384–5 record the names of 125 citizens who were assessed to pay for 52 new guns: 8 of twenty-four inches in diameter, 1 of twenty-one inches, 20 of twenty inches, 18 of eighteen inches, 3 of sixteen inches, 1 of twelve inches and 1 unspecified.[20] King's Lynn similarly made an assessment of its burgesses for this purpose on 21 August 1386, although the town possessed these weapons before this date, as the accounts for

14 KHLC, *Lydd Borough and Town Council*; LY/2/1/1/1, f. 1r.

15 KHLC, LY/2/1/1/1, ff. 4r, 17v, 22r, 61v, 64r.

16 KHLC, *Lydd Borough and Town Council.*

17 Richard Howlett, 'Norwich Artillery in the Fourteenth Century', *Norfolk Archaeology*, XVI (1907), 46–75 (pp. 60–1).

18 For the deployment of guns in royal castles see T. F. Tout, 'Firearms in England in the Fourteenth Century', *The English Historical Review*, 26 (1911), 666–702 (pp. 692–9).

19 William Hudson, ed., *The Records of the City of Norwich*, volume 1 (London: Jarrold & Sons, 1906), p. 396.

20 NRO, NCR case 18a/1, f. 4r; Howlett, 'Norwich Artillery in the Fourteenth Century', pp. 63–6.

1384–5 list a barrel of saltpetre for the use of guns.[21] The bailiffs of Canterbury, on 25 April 1385, were ordered to array all capable laymen to defend the city and to place guns and other weapons upon and around the walls.[22] The financial accounts for Canterbury do not state how many of these weapons were possessed by the city, but repairs to guns were carried out in the year 1396–7 and in 1403–4 they were moved from the Westgate to the 'Chamber'.[23] In addition, it is likely that many more towns had guns for which evidence no longer survives. For instance a 1377 petition from the burgesses of Dartmouth to the King and Council refers to the purchase of 6 guns, 10 'engines' and 1 chain for the price of £6.[24] There is also limited evidence that towns made use of their guns for the ships which they provided for naval expeditions. The town of New Romney acquired 2 guns from a Jacob Hikeman in 1400–1 by distraint, and spent 5s. on gunpowder for Henry IV's expedition to Scotland of the same year.[25] It is likely that other coastal towns in the south of England had guns by the beginning of the fifteenth century, as can be seen at the town of Hythe in 1412, where there is a reference to the movement of an unspecified number of guns.[26] Therefore the deployment of guns to royal castles appears to have been the factor which encouraged some towns to adopt gunpowder weapons in the 1380s.

CHRONOLOGICAL DEVELOPMENT

By contrast with the 1380s, the first half of the fifteenth century saw very limited expenditure by towns on guns.[27] This can be seen in the case of Canterbury and Lydd, which, despite having relevant surviving records, do not record any expenditure on guns before the 1450s. Instead it appears that the lack of expenditure was due to the increased security of the coastal regions of England following Henry V's conquest of Normandy.[28] This meant that only the far north continued to be threatened in this period by Scottish attacks, as can be seen by the dispatch of artillery to the

21 R. F. Isaacson, ed., *The Red Register of King's Lynn*, volume 2 (King's Lynn: Thew & Son, 1919–22), p. 27; KLBA, KL/C39/50.

22 *CPR 1381–1385*, p. 597.

23 CCA, CC/F/A1, ff. 14v, 30r.

24 TNA, SC 8/105, no. 5213.

25 KHLC, NR/FAC2, ff. 51r, 63v.

26 *Fourth Report of the Royal Commission on Historical Manuscripts* (London: H. M. S. O., 1874), p. 434. Unfortunately the Jurats' Book of Hythe appears to have been lost since it was unsatisfactorily calendared by the Fourth Report of the Historical Manuscript Commission in 1874.

27 See Appendix E.

28 C. T. Allmand, *Lancastrian Normandy 1415–1450: The History of a Medieval Occupation* (Oxford: Clarendon Press, 1983), p. 231.

royal castle at Berwick in 1434.[29] Activity was limited to the transport of gunstones in Southampton in 1433–4, as well as the purchase of gunpowder and repairs of guns in Dover from 1438 to 1440.[30]

This changed in the 1450s due to a dramatic increase in the amounts expended by the towns on gunpowder weapons, undoubtedly as a reaction to the loss of Lancastrian Normandy in 1450 and the renewed threat of French invasion.[31] This led to certain towns replenishing their stocks of artillery, such as Southampton, which spent heavily on repairing guns and gunpowder in the year 1449–50, and King's Lynn whose burgesses expended the sizeable sum of £1 6s. 8d. on the construction of new guns in 1451–2.[32] This also led other towns, some apparently for the first time, to spend money on firearms for their own defence, such as Bristol, which paid £40 for a dozen guns and saltpetre in 1450.[33] Rye paid for the carriage of two guns from Winchelsea in 1450–1 and in 1454–5 spent further sums on gunpowder.[34] Likewise, Canterbury spent £2 6s. 8d. on the purchase of a gun in 1449–50 and a further 2s. on its guns in the following year.[35] Even inland towns, such as Launceston, spent £3 1s. 1d. on the acquisition of guns in 1450, whereas Coventry allocated £6 5s. 7d. for four new guns in the following year.[36]

The greatest activity occurred in the years 1456–7 when the threat to England intensified. In 1457 a French fleet attacked Southampton but withdrew after being fired upon by the guns of the town. The fleet then attacked Sandwich in Kent and, despite the presence of a bulwark, was able to capture the town after heavy fighting.[37] This raid caused shock and alarm in the coastal communities of East Sussex and Kent, as can be seen by the sizeable sums they spent on acquiring guns in the following years: New Romney paid 6s. 8d. for two guns, Rye £2 13s. 4d. for six guns, Canterbury £2 13s. 4d. on one gun and Lydd spent the very large sum

29 TNA, E 404/50, no. 347.

30 Harry W. Gidden, ed., *The Steward's Books of Southampton*, volume 1 (Southampton: Cox & Sharland, 1935), pp. 53, 111; BL, Add MS 29810, ff. 19r, 28v.

31 See Appendix E. Ralph A. Griffiths, *The Reign of King Henry VI: The Exercise of Royal Authority, 1422–1461* (London: Benn, 1981), p. 428.

32 SRO, SC 5/1/7; KLBA, KL/C39/57.

33 E. W. W. Veale, ed., *The Great Red Book of Bristol*, volume 1 (Bristol: Bristol Record Society, 1931–1953), p. 131.

34 ESRO, RYE/60/2, ff. 24r, 42r–42v, 43v.

35 CCA, CC/F/A2, ff. 29v, 34r.

36 Richard Peter and Otho Bathurst Peter, *The Histories of Launceston and Dunheved, in the County of Cornwall* (Plymouth: Brendon & Son, 1885), pp. 133–4; Mary Dormer, ed., *The Coventry Leet Book, or, Mayor's Register, Containing the Records of the City Court Leet or View of Frankpledge, A.D. 1420–1555, with Divers Other Matters* (Oxford: Oxford University Press, 1907–13), p. 260.

37 Helen Clarke, *Sandwich: The 'Completest Medieval Town in England': A Study of the Town and Port from its Origins to 1600* (Oxford: Oxbow, 2010), p. 147.

of £6 1s. 7d. on the construction of a great serpentine.[38] A detailed insight into defensive preparations at this time can be seen in the accounts for Lydd in 1457, which show that the townspeople spent heavily on gunstones and gunpowder and include multiple entries for the guns moved to and from the 'Nesse', which is likely to have been some form of coastal headland.[39] At least part of this expenditure occurred during the raid itself as can be seen by a payment of meat and drink for two carpenters 'for a day mendyng the gunys the Mu(n)day that the Frencshe men came on lande'.[40] Rye also expended sizeable sums of money on gunpowder, repairs to guns and ammunition in the same period.[41] In addition, other coastal regions of England were threatened by attack, as can be seen by a letter to Sir John Fastolf at Caister Castle in 1458, when he was said to have 'shot many guns' in its defence.[42]

Expenditure on guns continued in the 1460s, particularly in the coastal towns of south-east England, but it was at a greatly reduced level from the 1450s.[43] The most significant years were 1460 to 1461 with dynastic strife reflected in payments by towns on guns. New activity can be detected in the cities of Canterbury and Exeter, whose inhabitants expended money on repairing their guns in 1460–1, with similar repairs carried out at Dover and Bridport in 1465–6.[44] Lydd continued to spend money on guns throughout the course of the 1460s, with the large sum of £2 19s. 11d. allocated for this purpose in 1460–1, although expenditure was at a reduced level from the previous decade.[45] The burgesses of Sandwich, undoubtedly with the experience of the raid of 1457 fresh in their memory, built up their stocks of gunpowder in 1462–3 and 1465–6, and spent the large sum of £3 9s. 7d. in 1468–9 on the purchase of gunpowder, guns, and for the keeping of the guns in the bulwarks.[46]

The 1470s saw an increase in expenditure on gunpowder weapons by the towns; this was due in part to the dynastic turbulence of the years 1470–1, as well as to war with France later in the decade.[47] In the early 1470s the towns of Kent

38 See Appendix E. KHLC, NR/FAC3, f. 47v; ESRO, RYE/60/2, f. 53v; CCA, CC/F/A2, f.56r; KHLC, LY/2/1/1/1, f. 50r.

39 KHLC, LY/2/1/1/1, ff. 47r–47v, 50r, 51r, 52r, 53v, 54r, 56r–56v, 57v–58r.

40 Ibid., f. 52r.

41 ESRO, RYE/60/2, ff.53v, 58v–60r.

42 John Fenn, ed., *Original Letters, Written During the Reigns of Henry VI, Edward IV, and Richard III* (London: G. G. J. and J. Robinson, 1787), pp. 156–7.

43 See Appendix E.

44 CCA, CC/F/A2, f. 14v; DHC, *Receivers' Roll Henry VI 39–Edward IV 1 1460–1461*; BL, Add MS 29616, f. 10r; *Sixth Report of the Royal Commission on Historical Manuscripts* (H. M. S. O., 1879), p. 494.

45 KHLC LY/2/1/1/1, ff. 61v–62r, 63Ar–63Av, 64r, 67r, 68v, 76v, 77v, 104r, 105r, 108r, 110r–111r, 112r–112v.

46 BL, Add MS 33511, ff.7v–8v; KHLC, SA/FAT5, f. 45r; SA/FAT6, ff. 16r, 17r.

47 See Appendix E.

incurred military expenses in supporting Richard Neville, earl of Warwick, during the Readeption of Henry VI, as can be seen in the case of Dover, which spent sizeable sums of money on the movement of guns, gunpowder and repairs.[48] In May 1471, the citizens of London used their artillery to successfully resist a four-day siege by an army led by the Bastard of Fauconberg, a supporter of Warwick, with subsequent repairs to their guns costing £15 19s. 2d.[49] Edward IV's success in regaining his throne this year led to the resumption of war with France, which prompted many towns to repair their existing stock of guns and to purchase gunpowder, notably in 1474–5, with Southampton spending £5 14s. 10d. and Dover £2 5s. 11d., in addition to smaller amounts expended by York, Lydd and New Romney.[50] Further expenses were also incurred by Dover later in the decade, with £3 18s. 7 ½d. spent on repairing the town's guns in 1477–8.[51]

The highest levels of recorded expenditure, however, occurred during the 1480s, due to tensions with France and dynastic strife following the death of Edward IV.[52] The years 1482 to 1484, in particular, were marked by heavy expenditure on gunpowder, with Dover spending £5 10s. 5d. in these years, Canterbury £4 10s. 1d., Rye £4 4s. 2d., Southampton £1 16s. 10d., and Lydd £1 7s. 2d.[53] As the price of gunpowder was largely stable from the 1450s to 1480s, this would indicate that these towns felt sufficiently threatened to require much larger quantities of gunpowder than in previous decades.[54] Sandwich continued to invest in firearms with its large arsenal of weaponry greatly enhanced by guns taken from captured French and Spanish ships; the burgesses of Rye also spent money repairing their bulwark and guns.[55] In the following year the city of Winchester contributed to its stock of gunpowder through the purchase of 20lbs for 12s. 2d.[56] The earliest

48 James E. Winston, 'English Towns in the Wars of the Roses' (unpublished doctoral thesis, University of Pennsylvania, 1921), pp. 55–7; BL, Add MS 29616, ff. 64r–65v, 67v, 83r, 86r; Egerton MS 2090, ff. 122r, 124v.

49 Charles Welch, *History of the Tower Bridge* (London: Smith, Elder and Co., 1894), pp. 262–3; LMA, COL/CC/01/01/007 f. 230r; COL/CC/01/01/008, f.26r; C. F. Richmond, 'Fauconberg's Kentish Rising of May 1471', *The English Historical Review*, 85 (1970), 673–92.

50 BL, Add MS 29616, f. 132v; Egerton MS 2090, f. 136r; R. B. Dobson, ed., *York City Chamberlain's Account Rolls 1396–1500* (Durham: Surtees Society, 1980), pp. 150, 153–5; KHLC, LY/2/1/1/1, f. 152v; NR/FAC3, f. 78r.

51 BL, MS Add 29616, ff. 152r–152v.

52 See Appendix E; C. S. L. Davies, 'Richard III, Brittany, and Henry Tudor 1483–1485', *Nottingham Medieval Studies*, 37 (1993), 110–26 (pp. 114–15).

53 BL, MS Add 29616, ff. 240v, 241r; CCA, CC/F/A/7, f.10; ESRO, RYE/60/3, ff. 16r, 82v; SRO, SC 5/1/18, f. 26r; SC 5/1/19 f. 46r; KHLC, LY/2/1/1/1, ff. 168r, 172r, 175v.

54 See Appendix D.

55 ESRO, RYE/60/3, ff. 46r–46v, 47v, 50r, 79v, 82v, 88v.

56 HRO, W/E1/35.

recorded payments for Plymouth also occur in 1486 with money being expended on lead shot for guns and for the keeping of the bulwark.[57] Dover continued with annual payments on its firearms for the remainder of the 1480s, on gunpowder, the movement of guns and repairs.[58] Sandwich also paid £2 2s. 9d. on repairing its guns in 1489–90.[59]

The 1490s saw a slight drop in the total amount expended, the vast majority of which was spent by the city of Exeter on guns, gunners and gunpowder during the siege of 1497.[60] War with France in the period 1489–92, however, also led some towns to carry out repairs to their arsenals, such as New Romney in 1490 and Sandwich in 1490–1, with Rye spending £1 1s. 8d. on the purchase of two guns in 1492–3.[61] The aldermen of Dartmouth also spent sizeable sums of money on furnishing their defences with guns, gunpowder and ammunition.[62] Heavy costs were borne by the city of Exeter in 1496–7 during the siege of the city by Cornish rebels which totalled the very large sum of £15 19s. 10 ½d., including £10 17s. spent on gunpowder.[63]

In summary, guns were first adopted by the towns in the 1380s, but their use went into abeyance until the 1450s; they were used at a low level for the next two decades, reaching a peak in expenditure in the 1480s, followed by a slight decline in the 1490s. The available sources show that the coastal towns of the south-east of England, particularly the Cinque Ports, were responsible for most of the recorded expenditure for the second half of the fifteenth century. New evidence from the 1480s onwards suggests that the towns in the south-west of England, principally Dartmouth, Exeter and Plymouth, were beginning to spend sizeable sums of money on gunpowder weapons by the beginning of the sixteenth century. Furthermore, although some inland settlements did spend money on guns, such as Coventry and York, the technology was mainly adopted by the coastal towns of England.

NUMBERS AND TYPES OF GUN

The financial evidence shows that considerable sums of money were spent by the towns on guns, but tracing the total number of weapons owned by them is difficult

57 PWDRO, W 130, ff. 5v–6r.

58 BL, Egerton MS 2107, ff. 4r, 10r, 19v, 20r, 29r, 30r; Add 29617, ff. 10v, 11r, 12r, 21r, 22r, 47v–48r.

59 KHLC, SA/FAT/9, ff. 5r, 8r.

60 See Appendix E.

61 KHLC, NR/FAC3, f. 102v; SA/FAT10, f. 1r; ESRO, RYE/60/3, f. 113r.

62 DHC, DD 61199a.

63 Mark Stoyle, *Circled with Stone: Exeter's City Walls, 1485–1660* (Exeter: University of Exeter Press, 2003), p. 185.

due to the small number of surviving inventories. The available evidence consists of one inventory for Sandwich (1482–3) together with three inventories and a letter for Southampton (1434, 1467–8, 1498–9, c. 1460) (see Figures 21, 22 and 23). The information for Southampton shows a significant increase in the number of guns owned between the beginning and the middle of the fifteenth century, with a decline in numbers occurring at the end of the century. The inventory for Sandwich also shows that at least some towns had large arsenals, with the town possessing fifty-three guns and four handguns in 1482–3; a comparable number to the sixty-five guns and eleven handguns held in the castle of Guînes in the Pale of Calais during the same year.[64] For most towns, however, this type of evidence does not exist and it is simply not possible to know how many guns they possessed. An example of this can be seen with the town of Lydd which in 1459–60 paid a John 'at Wyke' 14d. for repairing the guns, but does not specify the numbers involved.[65] Some idea of the number of guns held by the towns can be collected by adding up specific references to guns repaired, moved and purchased within individual years. This method shows that Norwich had at least fifty-two guns in 1384–5, Coventry four guns in 1451–2, as well as at least three guns and three handguns in 1471–2. Likewise, Lydd had four or more in 1458–9, Rye nine in 1483–3 and New Romney five in 1490–1. These numbers are likely to be quite misleading, however, as is revealed by the single surviving inventory for Sandwich, which is the only evidence for the town having a considerable quantity of weapons. Yet the available evidence does suggest that some towns were greatly increasing their stock of artillery in the middle part of the fifteenth century.

Financial documentation shows that towns were in possession of a variety of different gunpowder weapons, but most references simply refer to these weapons as 'guns'. In the first half of the fifteenth century, fowlers gradually became the most commonly described gun in royal inventories, yet there is only one example of the use of the term 'fowler' in the extant town accounts, for Coventry in 1469, suggesting that this terminology was simply not adopted by the townsfolk, who continued to use the generic word 'gun'.[66] This may have been because fowlers were the main type of gun used by the towns so there was no need to distinguish them from other types; however this was not the case with another popular type of gun: the serpentine. The earliest evidence in town accounts for this weapon is from Coventry whose burgesses purchased two large bronze serpentines from Bristol in 1450–1, a year after the earliest record of this gun type in royal records.[67] The presence of serpentines in urban arsenals gradually increased over time; in 1456–7 the town

64 KHLC, SA/FAT8 dorse; E 101/198/13, ff. 2, 3.

65 KHLC, LY/2/1/1/1, f. 57v.

66 See Chapter 1 and Appendix J for the development of gun types; Dormer, *The Coventry Leet Book*, p. 345.

67 Ibid., p. 260; TNA, E 364/87, rot. G.

of Lydd spent the sum of £6 1s. 7d. on the purchase of a large serpentine whereas the town of Rye had at least one serpentine in 1457–8 and multiple serpentines by 1482–3.[68] One of the highest numbers of recorded serpentines was at Sandwich, which had seven by 1482–3, out of a total of fifty-three guns.[69] Limited evidence exists for other types of guns possessed by the towns, with Southampton acquiring organ guns by 1449–50 and a large gun called *Thomas with the Beard* by 1467–8, whereas Sandwich had one murderer in 1482–3.[70] Later in the century, between 1487 and 1492, the aldermen of Dartmouth invested in six new murderers and sixteen serpentines to complement their existing arsenal, which already included two murderers.[71]

Small quantities of handguns were also stored in urban arsenals, with one handgun and one staff gun being kept in Coventry in 1470–1, four handguns at Sandwich in 1482–3, and an unspecified number at Southampton in 1498–9.[72] Some idea of the changes in the types of guns owned by the towns can be seen by comparing the inventories for Southampton and Sandwich. In 1467–8 Southampton had one serpentine out of a total of thirty-seven guns (2.7 per cent), whereas Sandwich had seven serpentines out of a total of fifty-three guns in 1482–3 (13.2 per cent). This suggests that serpentines were gradually becoming more important in town arsenals throughout the course of the latter half of the fifteenth century, a trend that is also reflected in royal accounts of the period.[73] It is possible that serpentines became increasingly popular in coastal towns as they could serve as anti-ship weapons. Their long thin shapes meant that they presumably could fire shot at greater distances than other types of guns.

DEPLOYMENT OF GUNS

The financial accounts show that towns often stored their gunpowder weaponry in civic buildings, such as court halls and council chambers. Canterbury kept its guns in the City Chamber in 1403–4 and 1460–1, although by 1470–1 these were stored in the Court House.[74] Likewise York stored guns in the Guildhall in 1468–9 and

68 KHLC, LY/2/1/1/1, f. 50r; ESRO, RYE/60/2, f. 59v; RYE/60/3, f. 50r.

69 KHLC, SA/FAT8 dorse.

70 See Appendix J and the Glossary for a definition of a murderer; SRO, SC 5/1/7, ff. 16r, 19v; KHLC, SA/FAT8 dorse.

71 DHC, DD 61199a.

72 Dormer, *The Coventry Leet Book*, p. 363; KHLC, SA/FAT8 dorse; Cheryl Butler, ed., *The Book of Fines: The Annual Accounts of the Mayors of Southampton*, volume 1, 1488–1540 (Southampton: University of Southampton, 2007), p. 2.

73 See Chapter 1 and Appendix J.

74 CCA, CC/F/A/1, f. 14v; CC/F/A/2, ff. 14v, 42r.

New Romney in the Common House in 1475–6.[75] Churches were also important storage locations for guns, as can be seen with Rye, which deposited its newly purchased guns in a church in 1456–7.[76] Dover also expended money on moving guns into St Martin's Church in 1486–7 and again in 1495–6.[77]

The best available evidence for how a town stored its arsenal of weaponry comes from the inventory for Sandwich in 1482–3, which demonstrates that more than half of the fifty-three guns (twenty-nine) were kept 'below the bulwark', comprising seventeen guns, ten great guns and two 'wasted guns' (see Figures 21 and 22). The heavy concentration of artillery at the bulwark shows that it was the principal defensive structure of the town. Another important location was the Court Hall which seemed to serve as a general storage location for the town, with thirteen guns kept there, comprising of two serpentines, six guns in three carts, together with a further five guns. Smaller quantities of ordnance were kept in other places, such as the castle (four), the Tower of St Davis (two), the adjoining tower on the left (two), immediately outside the bulwark (one), and elsewhere (two). The Council Chamber was used for storing gunpowder and 'the house above the bulwark' was used for keeping a variety of equipment including five crossbows, four handguns, eleven long pavises, thirty-one spears, thirteen brigandines, fifteen sallets and more gunpowder.[78]

The surviving documents for Dover, Lydd and Rye provide evidence that towns deployed their artillery in a systematic and organised fashion. The Dover accounts include numerous entries for guns to be moved to and from the walls in the late 1470s to 1490s, by porters making use of 'the great ladder'.[79] Guns were also sometimes deployed outside the walls, as can be seen in 1470–1, when the great guns were moved from the street to 'the cliff'.[80] The only evidence that the town used its artillery occurs in 1479–80, when 'diverse persons' were paid for setting forth of the guns and a James Tinker was paid for shooting guns when French ships were in the area.[81] The town of Lydd, despite lacking walls, regularly deployed guns and gunpowder to and from the 'Nesse' in 1458–9, 1468–9, 1469–70, 1471–2 and 1483–4.[82] A memorandum of 1470 specified that the masters of the ships were tasked with keeping an armed continual watch at the Nesse; the guards were to

75 Dobson, *York City Chamberlain's Account Rolls 1396–1500*, p. 130; KHLC, NR/FAC3, f. 78r.

76 ESRO, RYE/60/2, ff. 53v–54r.

77 BL, Egerton MS 2107, ff. 21r, 50r.

78 KHLC, SA/FAT/8 dorse.

79 BL, Egerton MS 2090, f. 152r; Add MS 29616, f. 64v, 192r, 208r; Add MS 29617, ff. 11r, 12r, 21r, 22r; Egerton MS 2107, ff. 4r, 47v, 201r.

80 BL, Add MS 29616, f. 65r.

81 Ibid., ff. 208r–208v.

82 KHLC, LY/2/1/1/1, ff. 47v, 51r, 52r, 53v, 54r, 104r, 110v, 119r, 175r–175v.

FIGURE 21 Types of guns in Sandwich inventory of 1482–3, excluding handguns.

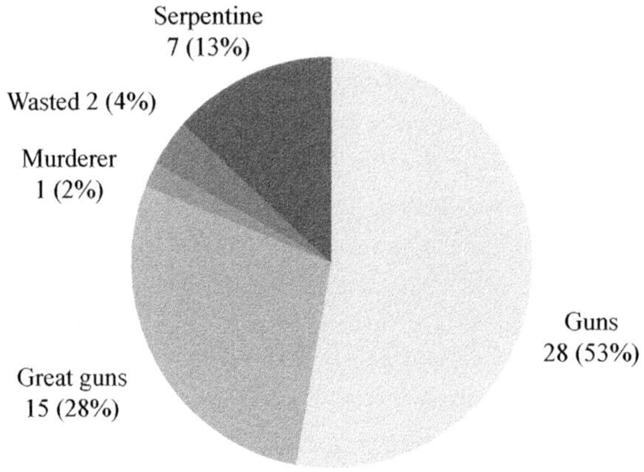

Source: KHLC, SA/FAT8 dorse.

FIGURE 22 Distribution of guns in Sandwich inventory of 1482–3, excluding handguns.

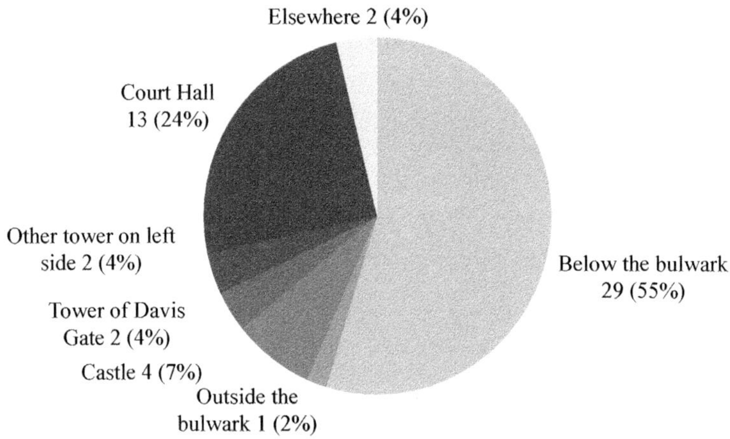

Source: KHLC, SA/FAT8 dorse.

be 'honest indwellers' (local residents) who should be equipped with at least four bows and four bills or staves, and they were to maintain the guns kept in the same location.[83] The stationing of artillery on this coastal headland is likely to have been motivated by the desire to provide a visual deterrent to enemy ships. Efforts were also made to gather intelligence on hostile shipping, with watchmen being regularly employed to look out for French and Danish ships at the Nesse, 'Ways End' and in the steeple of the church.[84] The town of Rye deployed its guns in 1457–8 and 1483–4 to gunports set in the walls of the town, with holes dug behind the walls into which posts would be set for holding the guns.[85]

Other towns also provide evidence of how they used their guns in particular periods of crisis. Exeter moved guns to its city gates during the siege of the city by the Cornish in 1497, as did Coventry when it was threatened in 1471.[86] In Hull, by contrast, the town's guns were positioned on the wharfs or given to all men 'that be of power' on the landside for the defence of the walls, when the inhabitants were menaced with attack by a Yorkist army in 1460.[87] Similarly in 1497, guns were deployed to defend the walls of Bristol from the Cornish rebels, with further ordnance used to equip ships and boats to protect the city by water.[88] The city of London was also threatened by rival armies during the Wars of the Roses, particularly in the early 1460s and 1470s. In 1460, the approach of the army of the Yorkist lords prompted a discussion where it was proposed by a pro-Lancastrian faction of the inhabitants that their entry should be blocked by placing guns on London Bridge. Subsequently, after admitting the Yorkists into the city, the Londoners laid siege to the Tower of London. According to the author of *An English Chronicle*, the Lancastrian garrison of the Tower fired small guns at the inhabitants, inflicting injuries on men, women and children in the streets. In response the Londoners 'leyde grete bombardes on the ferther syde of the Thamyse agayns the toure and crased the walles therof in diuerse places.'[89] Later, in 1470, after the flight of Edward IV from England in September, the citizens decided that all the gates should be guarded with armed men and guns, and a vigil held in all of the wards. The defence of London Bridge was attended to in the following month, with four gunners employed for three days and four nights, to oversee guns

83 KHLC, LY/2/1/1/1, f. 140r.

84 Ibid., ff. 47v, 48v, 51v, 54r, 61r–61v, 63v.

85 ESRO, RYE/60/2, f. 59r; RYE/60/3, f. 46r.

86 Stoyle, *Circled with Stone*, p. 185; Dormer, *The Coventry Leet Book*, p. 363.

87 HHC, C BRB/1, ff. 72–6.

88 L. T., Smith, ed., *The Maire of Bristowe is Kalendar, by Robert Ricart, Town Clerk of Bristol 18 Edward IV [1479]* (London: Camden Society, 1872), p. 49.

89 John Sylvester Davies, ed., *An English Chronicle of the Reigns of Richard II, Henry IV, Henry V, and Henry VI, Written Before the Year 1471* (London: Camden Society, 1856), pp. 94–6.

sent from the city's arsenal.[90] In May 1471 the city was besieged by an army and fleet led by the Bastard of Fauconberg, and the citizens were able to successfully resist assaults launched over the course of four days, in part due to the effective use of their guns. The defensive preparations undertaken by the defenders included fortifying the bank of the River Thames between Castle Baynard and the Tower of London. London Bridge was furnished with artillery supplied from the Guildhall, holes were cut into the drawbridge for the firing of guns, and it was strengthened with twelve 'sakkes' of stone and wool to protect it from gunshot.[91]

A lack of information means that it is generally difficult to ascertain how many guns were allocated to specific parts of towns. For Norwich, however, a comparison can be made between the list of men assessed to provide guns in the year 1384–5, and a list for the same year of men who were guardians of the gates and towers of the city.[92] This shows that the guardians of the eleven gates of the city were all liable to pay for at least one gun, as well as the guardians of twenty-one of the forty towers of the city. This would indicate that guns were concentrated in the defence of gates and, to a lesser extent, the towers. Therefore the towns often positioned their artillery within existing masonry defences, although the example of Lydd shows that un-walled towns could also make use of these weapons in other ways.

ARTILLERY FORTIFICATIONS

The earliest gunport in England is said to be at Quarr Abbey in the Isle of Wight in 1365 and the construction of these features proliferated in the 1370s and 1380s, as can be seen at the West Gate of Canterbury, constructed in 1380, which included twenty gunports.[93] The presence of gunports in extant town walls, such as at Southampton (c.1360–80), indicates that more towns possessed guns in the late fourteenth century than the surviving documentary evidence suggests.[94] This is also supported by the evidence from Norwich, which shows that the town had the substantial quantity of fifty-two guns to defend its walls and artillery tower.[95] Gunports are rarely mentioned in the extant town accounts, except at Southampton,

90 LMA, COL/CC/01/01/007, f. 222r; Welch, *History of the Tower Bridge*, p. 262.

91 Ibid., pp. 262–3; LMA, COL/CC/01/01/007, f. 230r; COL/CC/01/01/008, f. 26r.

92 Howlett, 'Norwich Artillery in the Fourteenth Century', pp. 63–74.

93 D. F. Renn, 'The Earliest Gunports in Britain?' *Archaeological Journal*, 125 (1968), 301–3; John R. Kenyon, 'Early Artillery Fortifications in England and Wales: A Preliminary Survey and Reappraisal', *Archaeological Journal*, 138 (1981), 205–40 (pp. 207–12); B. H. St J. O'Neil, *Castles and Cannon: A Study of Early Artillery Fortifications in England* (Oxford: Clarendon Press, 1960), pp. 7–8.

94 Kenyon, 'Early Artillery Fortifications in England and Wales', p. 210.

95 NRO, NCR case 18a/1, f. 4r.

in 1449–50 and 1456–7, when men were paid for making and repairing gunports, and Rye in 1457–8, when fresh gunports were made in the walls of the town.[96] Another form of defensive architecture employed in town defences was the artillery tower. The earliest known example of this type of fortification is the Cow Tower in Norwich, constructed in 1398–9, which consists of a three-storey structure outside the circuit of the city walls, with the upper two levels containing two tiers of gunports in a circuit around the tower.[97] Another artillery tower from this period is God's House Tower in Southampton, built early in the fifteenth century on the south-eastern corner of the town, which was used to store a sizeable proportion of the town's artillery.[98]

The most significant development in artillery fortification in the fifteenth century was the advent of the bulwark.[99] These structures were perhaps first encountered by the English in their siege of Harfleur in 1415, where the bulwarks were subsequently repaired and added to by the garrison.[100] The earliest recorded reference for the use of these buildings in England occurs in 1420–2, when £33 4s. 7d. was expended on the construction of a 'wooden tower called a bulwark' on the River Hamble to protect the king's ships.[101] These structures were not adopted by the towns, however, until the invasion scares of the 1450s. In 1451 Sandwich built a fortification for guns, originally called the 'new wall', later 'the bulwark', in the north-eastern corner of the town, although it was overrun when the town was attacked by the French in 1457.[102] The following year, the Corporation of Sandwich decided to build a new bulwark of brick besides Fisher's Gate 'in all hast', which served as an important location for the positioning of the town's guns in the 1480.[103] It continued to serve as the principal defensive feature of the town in the following decade, as can be seen from a memorandum of 13 June 1492 which stated that 'all … such gunnys as lately wer broken and new beyng in the bulwek shalbe by the treserere stockid and amendid sufficiently for defence of the towne'.[104]

The raid of 1457 was also important in encouraging this development in other towns in England, who record the use of this type of fortification for the

96 SRO, SC 5/1/7, f. 16v; SC 5/1/8, f. 12v; SC 5/1/11, f. 12r; ESRO, RYE/60/2, f. 59r.

97 A. Saunders, 'The Cow Tower, Norwich: An East Anglican Bastille?', *Medieval Archaeology*, XXIX (1985), 109–19 (pp. 109, 115).

98 Moffett, 'The Military Organization of Southampton in the Late Medieval Period', p. 43. For more details see Chapter 7.

99 See Glossary for definition. For a contrary view regarding the presence of bulwarks in late medieval England see Kelly DeVries, 'Gunpowder Weaponry and the Rise of the Early Modern State', *War in History*, 5 (1998), 127–45.

100 TNA, E 36/79, ff. 40, 56–7.

101 TNA, E 364/61, rot. G dorse.

102 Clarke, *Sandwich*, pp. 147, 154.

103 KHLC, SA/AC/1, f. 105v.

104 KHLC, SA/AC2, f. 41r.

first time, such as Great Yarmouth, which was given money by Exchequer for the construction of a bulwark, and Southampton, which deployed guns to an already existing bulwark.[105] Other towns which constructed bulwarks include Rye, with one recorded in 1457–8, and another one built in 1483–4, as well as Dartmouth, whose inhabitants had begun construction of a bulwark by 1481, and Plymouth, which possessed a bulwark by 1486–7. A chance reference to the presence of a bulwark situated in front of the south-east gate of Bristol, during a visit by William Worcester to the city in 1480, suggests that they were also present at other settlements for which documentary evidence does not survive.[106] Financial accounts make clear that bulwarks were often constructed of timber and earth, although they could also be made of brickwork, such as at Rye in 1483–4, or stone as at Dartmouth in 1481.[107] These structures were situated in the most important parts of the town defences and were well endowed with artillery; this is made explicit in the indenture between Edward IV and the town of Dartmouth in 1481, which states that the new fortification should be 'garnished with guns and artillery and other ordnance'.[108] Bulwarks were therefore considered useful for the defence of English towns in the fifteenth century, although other towns which made significant use of guns in the fifteenth century, such as Dover and Lydd, provide no evidence of possessing bulwarks. English towns were therefore clearly willing to invest in new artillery fortifications to improve their defences and to utilise their stocks of gunpowder weapons.

PERSONNEL

The earliest known evidence for the employment of a gunner in a town was in 1386 when one gunner was employed on improving the defences of Rye.[109] Southampton was unusual in having a town gunner intermittently for much of the second half of the fifteenth century; instead most towns usually hired gunners for specific purposes, such as at Rye in 1456–7, when a gunner was paid to assess six new guns and in the following year paid for three gun chambers.[110]

105 TNA, E 364/93, rot. F; SRO, SC 5/1/8, f. 12r.

106 ESRO, RYE/60/2, ff. 32r, 58v; RYE/60/3, ff. 79v, 88v; *CPR 1476–1485*, p. 251; PWDRO, W 130, f. 5r; James Dallaway, ed., *Antiquities of Bristow in the Middle Centuries; including the Topography by William Wycestre, and the Life of William Canynges* (Bristol: Mirror Office, 1834), p. 149.

107 ESRO, RYE/60/3, f. 32R; *CPR 1476–1485*, p. 251.

108 *CPR 1476–1485*, p. 251.

109 TNA, E 364/22, rot. G dorse.

110 Southampton's gunners are discussed in more detail in Chapter 7. ESRO, RYE/60/2, ff. 53v, 59v.

Gunners were generally only hired on a temporary basis during periods of emergency, such as at London in 1470 and Coventry in 1471 when the communities were threatened by rival armies, or in Exeter, which in 1496–7 temporarily hired ten gunners to help defend the city against Cornish rebels.[111] No information exists about the backgrounds of these men, but it is clear that a certain number of these professionals were in existence in the late fifteenth century and could be hired by towns in periods of crisis. Few towns, therefore, appear to have made use of gunners before the sixteenth century, although some of them employed men known as keepers of the artillery or the town bulwarks. In 1468–9, a William Brykeman was listed as an officer of the town of Sandwich and paid £1 6s. 8d. for keeping of the bulwark and guns for a year.[112] By 1482–3 a George Baring was paid £1 for keeping of the bulwark for a year and 5s. for a gold cloth for his livery, with further payments being made to him for repairing guns in the bulwark in 1489–90 and 1490–1.[113] Plymouth also employed a keeper for its bulwark, William Berry, who was paid 6s. 8d. in 1486–7, and who received a pension of a 1s. a year for this in 1495–6.[114] York was another settlement which saw fit to appoint a keeper for its artillery, with a John Craven receiving £1 for his fee in addition to 4d. for his livery in 1475–6, £1 for his fee together with 6s. 8d. for his livery in 1478–9, and £1 for his fee and 11s. 8d. for his livery in 1486–7. On 29 April 1484, however, it was reported to the community of the city that John Craven had been deficient in carrying out his duty of maintaining the guns and had instead 'suffered them to rust and be lost' and so should lose his fee; an issue that was still unresolved by the corporation on 17 June 1488.[115] These individuals appear to have had many of the skills and responsibilities of gunners, as can be seen in the case of George Baring, who received payments for stocking and repairing guns in the 1480s.[116] Keepers of artillery, therefore, appear to have been relatively well paid, but were employed on an intermittent basis. It is generally unclear which individuals were responsible for looking after the ordnance in many towns, for example at Lydd and Dover, where we have no relevant information or names. Therefore it appears that the continuous employment of gunners or keepers of artillery was not considered necessary for the supervision of ordnance, even for towns with large arsenals such as Sandwich.

111 Dormer, *The Coventry Leet Book*, p. 366; Welch, *History of the Tower Bridge*, p. 262; Stoyle, *Circled with Stone*, p. 185.

112 KHLC, SA/FAT6, f. 16.

113 KHLC, SA/FAT8, m. 48; SA/FAT9, f.5r; SA/FAT10, f. 1r.

114 PWDRO, W 130, ff. 6r, 35r.

115 Lorraine C. Attreed, ed., *The York House Books: 1461–1490*, volume 2 (Gloucester: Alan Sutton for Richard III & Yorkist History Trust, 1991), pp. 307, 617.

116 KHLC, SA/FAT8, ff. 58r, 68r; SA/FAT9, f. 5r.

MOTIVATIONS

Expenditure on guns by the towns was primarily a reaction to defensive threats, as can be seen by the significant outlay in particularly tense periods such as 1457–8, 1482–4 and 1496–7.[117] At a meeting of the Corporation of Bristol, on 2 August 1450, the decision was taken to spend £40 of the communal money on the purchase of 'gonnes and other stuffe necessarie for defence of the said town'.[118] In the following year, the aldermen of Coventry, although inland, justified their purchase of four guns 'for strengthening of the same if need should it require, (the which God forbid!)'.[119] Likewise, on 21 January 1470, the bailiff and jurats of Lydd agreed that the masters of the ships should keep a continued watch at the Nesse, which included maintenance of the guns there.[120] Further evidence for this can be seen with a meeting of the representatives of the Cinque Ports at Dover on 6 February 1475, where it was ordained by the consent of 'all towns' to 'grant to bear the charge of reparation of guns, gunpowder, gunstones and other habiliments of war'.[121] Three years later, it was recorded in the accounts of Launceston that one of the obligations of the aldermen was to find men and guns when the king required it.[122] Some towns also appear to have had some conception of how many guns they should possess for their defence and how they should be employed. This can be seen from the letter from the Corporation of Southampton to the 'Lords of the Council' (c.1460), which stated that the town's walls were incapable of withstanding gunshot and only had twenty-five guns when sixty were required.[123] Town governors were also keen to ensure that their artillery could be used to best effect, as can be seen from the order issued by the Corporation of Sandwich on 1 June 1478, which ordained that any ships anchored besides the bulwark and Fisher's Gate should be moved, on pain of forfeiture, so that the town's guns should have 'playn wey of shetyng'.[124] Later, in 1484, the citizens of London ordered that a survey should be carried out of the city's artillery, following a fire, and any damaged guns or gunpowder replaced.[125]

117 See Appendix E.

118 Veale, *The Great Red Book of Bristol*, volume 1, p. 131.

119 Dormer, *The Coventry Leet Book*, p. 260.

120 KHLC, LY/2/1/1/1, f. 140r.

121 BL, Add MS 29616, f. 136r.

122 Peter and Peter, *The Histories of Launceston and Dunheved, in the County of Cornwall*, p. 160.

123 R. C., Anderson, ed., *Letters from the Fifteenth and Sixteenth Centuries: From the Archives of Southampton* (Southampton: Cox and Sharland, 1921), p. 21.

124 KHLC, SA/AC/1, f. 241r.

125 LMA, COL/CC/01/01/009, f. 43v.

The growing importance of these weapons is also reflected in the emergence of a new category of headings in town accounts, such as those for Canterbury in the year 1457–8, in which the section 'the upkeep of houses and mills' was changed to 'the upkeep of houses, mills and guns', although it was rare for guns to be mentioned in this way consistently year-by-year.[126] Additionally it appears that towns were keen to showcase their guns, particularly to the burgesses who had funded their construction and upkeep. For instance on 25 July 1386, a discussion occurred at Norwich concerning the guns to be made and how they 'shall be prepared for their exhibition to the citizens on the Feast of St. Peter ad Vincula'.[127]

The surviving evidence shows that purchases of guns by the towns mainly occurred in specific periods, primarily the 1380s and 1450s.[128] On occasion this is likely to have funded by special financial contributions levied from the community specifically for this purpose, as at King's Lynn and Norwich in 1386.[129] However, the available sources suggest that many of these settlements first acquired guns in the 1450s, as this was when they first started to expend money on these weapons.[130] For instance the earliest evidence of expenditure by Lydd occurs in 1456–7, which is also when the community spent the sum of £6 1s. 7d. on a large serpentine.[131] It is unclear exactly why the town chose to invest so heavily in a large gun, as opposed to a greater number of smaller weapons, but it suggests that the bigger guns were considered important for defence. The extant financial information does not provide a full explanation of how and when these towns acquired gunpowder weapons. For instance the earliest evidence for Sandwich is in 1458–9, when the town spent 8d. on gunpowder, yet the accounts first record purchases of guns in 1468–9.[132]

Guns were acquired from a variety of sources, such as from the burgesses themselves, as can be seen with the weaponry appropriated in New Romney in 1400–1 and Sandwich in 1490–1.[133] Alternatively they were donated by individuals, as at Salisbury in 1483, when a serpentine was given by a William Tavern for the defence of the city. They could also be acquired through acts of war, as when Sandwich was able to acquire a large quantity of artillery from captured French and Spanish ships in 1482–3.[134] The towns also shared weapons amongst themselves. In

[126] CCA, CC/F/A/2, f. 56r.

[127] Howlett, 'Norwich Artillery in the Fourteenth Century', p. 49.

[128] See Appendix E.

[129] R. F. Isaacson, ed., *The Red Register of King's Lynn*, volume 2 (King's Lynn: Thew & Son, 1919–22), p. 27; Howlett, 'Norwich Artillery in the Fourteenth Century', p. 49.

[130] See Appendix E.

[131] KHLC, LY/2/1/1/1, f. 50r.

[132] KHLC, SA/FAT3; SA/FAT6, m. 16.

[133] KHLC, NR/FAC2, f. 51r; SA/FAT10, m. 1r.

[134] WSRO, G 23/1/2, f. 148v; KHLC, SA/FAT8, m. 68r.

1450–1 Rye paid for two guns to be transported from Winchelsea and six guns from the same place in 1483–4. Later, in 1482–3, Sandwich sent a serpentine to the town of Deal by 'the mayor's commandment'.[135] These settlements were clearly in regular communication with each other. For instance Lydd in 1460–1 despatched men to Winchelsea as well as New Romney for tidings of Frenchmen, and delivered a letter to Rye to warn them of the same.[136] These links also extended to the purchase of gunpowder, with Lydd acquiring gunpowder from Sandwich in 1465–6 and Dover in 1469–70, whereas Dover obtained gunpowder from Sandwich in 1482–3 and Gravesend in 1488–9.[137] Therefore, it appears that expenditure on gunpowder weapons was often a response to specific external threats, with peaks in particular periods of crisis, such as in the 1380s, 1450s and 1480s.

ROYAL POLICY

Successive kings were keen to ensure that the towns of England were well defended from attack, and commissions of array were often issued for the defence of coastal regions and the towns, such as for Great Yarmouth in 1457.[138] Despite this, there is limited evidence of kings issuing commands to towns to make use of guns for their own defence.[139] Nevertheless, royal support was important for the construction and maintenance of artillery fortifications in the towns. English kings had long encouraged the construction of urban defences, for example, by grants of murage, but support for architecture especially adapted to gunpowder weapons was a new development.[140] Royal subsidies were provided for the construction of bulwarks, an example being the £20 provided to Great Yarmouth in 1457.[141] The following year the town spent a further £20 on purchasing four guns, 120lbs of gunpowder, and on carrying out further work on its fortifications.[142] In 1462, Dartmouth was granted £30 per annum for twenty years for the defence of the borough, including the use of chains, guns and gunpowder.[143] This money was evidently put to good use, as by 1481, the grant was extended for a further five years to fund the work already started for the

135 ESRO, RYE/60/2, f. 24r; RYE/60/3, f. 47v; KHLC, SA/FAT8 dorse.

136 KHLC, LY/2/1/1/1, ff. 63Br, 64r.

137 KHLC, LY/2/1/1/1, ff. 51r, 112v; BL, Add MS 29616, f. 241v; Add MS 29617, f. 48r.

138 *CPR 1452–1461*, p. 371.

139 One of the few exceptions to this can be seen in the case of Canterbury in 1385, when the bailiffs of the city were appointed to array all capable laymen and to cause guns and other engines to be prepared, *CPR 1381–1385*, p. 597.

140 Turner, *Town Defences in England and Wales*, p. 24.

141 TNA, E 364/93, rot. F.

142 TNA, E 364/98, rot. E.

143 *CPR 1461–1467*, p. 75.

construction of a strong tower and bulwark, which was to be furnished with guns.[144] The bulwark at Plymouth, in existence by 1486, may have been partly funded by the 100 marks a year allocated to the town for its walls, as granted by Richard III in 1483.[145] In 1461 Sandwich was granted £100 per annum towards the fortifications of the town in return for a contribution of £20. By 1465 this had been replaced by exemptions from the customs duties, with some of the accounts for this work surviving for the 1460s and 1470s.[146] Royal support for the construction of bulwarks, on occasion, even extended to the participation of royal officials, as at Southampton in 1470 when William Temple, clerk of the king's ordnance, oversaw the works in the town. In the same year the burgesses were given a discharge for money outstanding for the fee farm in part due to 'the charges that they by oure comaundement must bere in stokkyng of gunnes and makyng of bulwerks for the defense of oure saide town'.[147] Likewise in 1488, Sir William Tyler, was paid £40 towards the 'making of anewe bulwarke besides our towne and castell of Berwic for the defense and saufegard of the same'.[148] Later in the sixteenth century, the antiquarian John Leland claimed that Richard III had constructed a bulwark at Scarborough on the sea shore.[149]

A detailed series of accounts showing how the aldermen of Dartmouth spent their annuity of £40 survives for the reign of Henry VII between 1487 and 1494. Most of the payments relate to the repair of the fortifications of the port, primarily the castle, but there are also numerous entries relating to the ordnance. Over a five-year period, from 1487 to 1492, the town acquired twenty-two guns for the defence of the castle, consisting of two large murderers, four other murderers and sixteen serpentines; repairs were also carried out to two existing murderers. Further payments were made for the purchase of gunpowder and ammunition, the construction of a bulwark at Kingswear and for the wages of four watchmen employed to guard the castle and ordnance.[150] This activity coincided with war with France, from 1489 to 1492, which suggests that Dartmouth was considered to be vulnerable to French raids, and hence why royal subsidies were provided to augment its defences.[151] By contrast, no additional artillery was acquired in the final

144 *CPR 1476–1485*, p. 251.

145 Rosemary Horrox, ed., *British Library Harleian Manuscript 433. Vol. 1, Register of Grants for the Reigns of Edward V and Richard III* (Gloucester: A. Sutton for the Richard III Society, 1979), p. 168.

146 TNA, E 101/184/26, 27, 28, 29; E 101/184/26, 27, 28, 29.

147 SRO, SC 5/1/12, ff. 17r–17v; TNA, E 159/247, brevia directa baronibus, Trinity, rot. 12.

148 TNA, E 404/79, no. 156.

149 L. T. Smith, ed., *The Itinerary of John Leland in or About the Years 1535–1543. Parts I to III* (London: George Bell and Sons, 1907), p. 60.

150 Kingswear Castle is located on the other side of the River Dart; DHC, DD 61199a.

151 For the war with France see J. M. Currin, '"The King's Army into the Partes of Bretaigne": Henry VII and the Breton Wars, 1489–1491', *War in History*, 7 (2000), 379–412.

two years of the accounts, with £103 15s. 10d. instead spent on carrying out repairs to the fortifications.[152] These documents demonstrate that the financial support provided by the crown in the late fifteenth century could make a substantial difference to urban defences; by this means Dartmouth Castle was furnished with sizeable quantities of ordnance. The aldermen of Dartmouth were particularly fortunate to receive annual payments from successive English kings, as it meant that they could modernise and update their defences without incurring a heavy financial burden.[153] By contrast, the artillery fortifications constructed to protect Portsmouth were built directly under the auspices of royal officials. In the early fifteenth century, a clerk of the works, Robert Barbot, was tasked with building a tower for 'the safe keeping of the king's ships', later known as the Round Tower, which was equipped with three guns in 1422.[154] The defences of the settlement were later improved in the reign of Henry VII through the construction of a tower and blockhouse on the other side of the harbour entrance that was designed to protect the newly built royal dock.[155] An inventory of 1495 reveals that the blockhouse was equipped with five guns, including a large serpentine and two murderers.[156]

Royal guns were also issued to the towns for their own defence in the second half of the fifteenth century. The earliest known record of this occurred in 1450, when Sir Thomas Kyriell donated guns to Southampton prior to the departure of his army for France.[157] The government of Henry VI later responded to the French raid on Sandwich in 1457 by distributing its stores of artillery to threatened places, as can be seen by a request made in December 1458 for John Judde, master of the ordnance, to be paid for placing ordnance upon 'the see coostes'.[158] These measures included the despatch of supplies to towns, as can be seen by the eleven barrels of gunpowder, saltpetre and sulphur sent to Bristol at this time.[159] In the next year, following the battle of Ludford Bridge, Judd was ordered to seize the artillery of the fugitive Yorkist lords and to survey the condition of all the ordnance in castles and towns in England to ensure that they were ready for defence.[160] Three men were also appointed to purvey guns and gunpowder for the

152 DHC, DD 61199a.

153 Note that further subsidies were provided in the sixteenth century for Dartmouth, for example see DHC, DD 61249A.

154 H. M. Colvin, *The History of the King's Works*, volume 2 (London: H. M. S. O., 1963), pp. 792–3; TNA, E 101/479/25; E 364/66, rot. H.

155 Another tower, called the Square Tower, is also attributed to this period. Colvin, *The History of the King's Works*, volume 2, pp. 495–6.

156 M. Oppenheim, ed., *Naval Accounts and Inventories of the Reign of Henry VII, 1485–8 and 1495–7* (London: Naval Records Society, 1896), p. 338.

157 SRO, SC 5/1/7, f. 17v.

158 TNA, E 404/71/3, no. 43.

159 Veale, *The Great Red Book of Bristol*, volume 1, p. 136.

160 *CPR 1452–1461*, p. 527.

defence of Dover Castle and the defence of the adjacent ports in September 1457.[161] Later, in December 1459, Henry VI ordered that thirty guns with gunpowder and shot be delivered to the bailiff of Gloucester, 'for the seuretee and savegarde of oure saide towne'.[162] Edward IV also strengthened urban defences through the provision of guns to strategic towns. In 1470–1 the city of Canterbury was provided with a gun captured from the supporters of the Bastard of Fauconberg at Blackheath. Later, in 1477–8, the town accounts for Dover record 'costs and reparacions done upon the king's gunnes'. These weapons were transported from London to Sandwich before they were again moved to Dover, where they were repaired at the town's expense.[163] Similarly the city of Carlisle was sent four barrels of gunpowder together with one serpentine and one handgun in 1480.[164] This policy was continued by Edward's successors, as can be seen from Richard III's instruction, in March 1485, that the master of the ordnance should despatch a barrel of gunpowder and three carts of 'rennyng ordenance' for the defence of Harwich.[165] Likewise in 1490, Henry VII ordered that two large serpentines should be transported from the ship the *Saint Katherine* to Winchelsea for its defence, where they were repaired at the expense of the Exchequer.[166] In the same year, £73 6s. 8d. was spent on artillery for equipping Berwick and Carlisle, and further ordnance was purchased for Berwick, Scarborough and Newcastle the following year.[167] Later on 16 August 1491, Henry Wyot, overseer of the castle of Carlisle, transferred seven guns and three barrels of gunpowder to the mayor of the city for its defence.[168]

These royal contributions were likely to have proved crucial for the strengthening of urban defences, as the example of Lydd in 1456–7 demonstrates that guns could be very expensive.[169] This may also explain why proportionally less money was spent on the purchase of guns by the towns from the 1460s to the 1480s. The few relevant surviving letters sent to the government provide some indication into how this process occurred. On 15 March 1458, John Bokking, a member of Sir John Fastolf's household, wrote a letter to him, to advise that he had asked the Archbishop of Canterbury to 'move the council for refreshing of the town of

161 *CPR 1452–1461*, p. 401.

162 TNA, E 404/71/4, no. 24.

163 CCA, CC/F/A/2, ff. 41v–42r; BL, Egerton MS 2090, f. 152v.

164 Michael R. McCarthy, *Carlisle Castle: A Survey and Documentary History* (London: English Heritage, 1990), p. 162.

165 Rosemary Horrox, *British Library Harleian Manuscript 433. Vol. 2, Second Register of Richard III* (Gloucester: A. Sutton for the Richard III Society, 1980), p. 223.

166 TNA, E 404/80, no. 237; E 405/77, m. 6d.

167 TNA, E 405/78, mm. 28d, 31v, 36r.

168 McCarthy, *Carlisle Castle*, p. 163.

169 KHLC, LY/2/1/1/1, f. 50r.

Yarmouth with stuff of ordnance and guns and gunpowder'.[170] The Corporation of Southampton, perhaps encouraged by its donation of royal guns in 1450, wrote to the 'Lords of the Council' to request additional artillery in a letter dated circa 1460.[171] Similarly, the town of Sandwich sent a delegate to Richard III at London in 1483 to request ordnance as it was 'lakkying artillery and gonpowder to defend the town'.[172] Later, in 1487, the city of York, threatened by supporters of Lambert Simnel, wrote to Henry VII requesting assistance as 'youre said citie is soo gretely decayed aswell by falling downe of the walles ... and where also your saide citie is not well furnesshed with artilment and stuff of ordnaunce'.[173] Henry VII responded by stating that the city could have twelve serpentines from Scarborough Castle, in return for paying for the costs of their carriage, although the constable of the castle, William Tunstall, was ultimately unable to provide the artillery.[174] These examples suggest that there was a dialogue between the towns and crown, with requests from the former, on occasion, resulting in the despatch of artillery by the latter to important locations. By contrast, there is very little evidence for towns lending or giving guns to the crown, except at Southampton. Here on 30 May 1468 the town governors delivered by indenture a large gun called *Thomas with the Beard* to Lord Scales, presumably for the planned naval expedition to Brittany.[175]

CONCLUSION

Medieval towns in England first adopted guns in the 1380s, most likely inspired by the deployment of guns in royal castles, although their usage largely went into abeyance until the 1450s. Thereafter, investment in gunpowder weapons was considered crucial for the defence of towns, and particularly heavy expenditure occurred in the 1480s and 1490s.[176] The pattern of firearms expenditure generally reflects the perceived threats posed to English towns, principally from French raiders or invaders, as in 1450 and 1457. The data also indicate that there was some regional variation in the adoption of this new technology, with the Cinque Ports, in particular, making extensive use of firearms for their defence from the 1450s until the early 1480s, although the majority of the recorded expenditure for the

170 Fenn, *Original Letters, Written During the Reigns of Henry VI, Edward IV, and Richard III*, pp. 156–7.

171 Anderson, *Letters from the Fifteenth and Sixteenth Centuries*, p. 21.

172 KHLC, SA/AC/1, f. 292r.

173 Attreed, *The York House Books*, volume 2, p. 549.

174 Ibid., pp. 556, 562.

175 SRO, SC 5/1/11, f. 12r; Scofield, *The Life and Reign of Edward the Fourth*, volume 1, p. 472.

176 See Appendix E.

remainder of the century was undertaken by the towns of the south-west, namely Dartmouth, Exeter and Plymouth. The substantial gaps in the surviving sources mean that this assessment needs to be treated with caution, although it is beyond doubt that guns were a key weapon in the arsenals of many coastal communities in southern England. The evidence also demonstrates that the aldermen of these settlements were able to keep abreast of technological changes, whether it was new types of guns, such as serpentines, or developments in artillery fortifications, notably bulwarks. New ideas and innovations seem to have spread quickly across different regions of England, perhaps in part due to the close links between many of the urban communities of the south coast and Continental ports. It is therefore surprising that the majority of towns that possessed guns did not, or only infrequently, employ gunners or keepers of artillery to maintain and organise their ordnance. This suggests that most urban authorities assumed that they could generally deploy and repair their artillery without specialist assistance, with the services of gunners, on the whole, only considered essential in desperate circumstances, such as during the siege of Exeter in 1497.

A significant reason for these developments was the support provided to the towns by the governments of Henry VI, Edward IV and Henry VII. Successive English kings were prepared to finance the construction of new artillery defences, particularly bulwarks, and even to send royal guns to supplement urban arsenals. This has important implications for previous interpretations of royal policy towards the possession of gunpowder weapons by the towns. The argument that the mid-fifteenth century was a period of decline as it saw a loss of control by the crown over gunpowder weapons can be shown to be inaccurate.[177] On the contrary, English kings, rather than seeking to restrict the supply of weaponry available to the towns, actively supported them in acquiring new guns. The importance of the Tudor monarchs in providing firearms for the towns, as identified by previous scholars, can also be partly reassessed, as this process instead began in the mid-fifteenth century.[178] Therefore, the development of gunpowder weapons in English towns was due to a successful collaboration between the central government and urban authorities.

177 DeVries, 'Gunpowder Weaponry and the Rise of the Early Modern State', pp. 142–3.

178 Steven Gunn, David Grummitt and Hans Cools, *War, State and Society in England and the Netherlands, 1477–1559* (Oxford: Oxford University Press, 2006), pp. 64–5.

7

Southampton

It is difficult to trace the development of gunpowder weapons in most individual towns due to the fragmentary nature of the extant evidence and a lack of inventories. Fortunately, it is possible to carry out an in-depth study of the use of guns by the town governors of Southampton, due to a relative abundance of surviving sources. This chapter demonstrates that the town's experience of firearms, whilst sharing some similarities with other coastal settlements, was unique. By the middle of the fifteenth century, Southampton had become one of the most prosperous towns in England, largely as the result of having secured two staples, wool and metals, which were exported from the town by Italian merchants.[1] This trade declined in the later years of the fifteenth century, with a fall in custom revenues in the early years of the reign of Edward IV, but it revived and reached a peak under the reign of Henry VII.[2] The period also saw significant changes to the government of the community, with a charter of incorporation in 1445 and the creation of the county of the town of Southampton in 1447.[3] These charters, and later ones, established the right of the town to self-governance and freedom from the interference of the county of Southampton or of the sheriff.[4]

The town was also important due to its military role in the late medieval period. It operated as a point of embarkation for many English armies transported to France and provided ships for naval expeditions.[5] This meant that it was vulnerable

1 James E. Winston, 'English Towns in the Wars of the Roses' (unpublished doctoral thesis, University of Pennsylvania, 1921), p. 33; Cheryl Butler, ed., *The Book of Fines: The Annual Accounts of the Mayors of Southampton. Vol.1, 1488–1540* (Southampton: University of Southampton, 2007), p. viii; Alwyn A. Ruddock, *Italian Merchants and Shipping in Southampton, 1270–1600* (Southampton: University College, 1951), p. 68.

2 Ruddock, *Italian Merchants and Shipping in Southampton*, p. 258.

3 Colin Platt, *Medieval Southampton: The Port and Trading Community, A.D. 1000–1600* (London: Routledge and Kegan Paul, 1973), p. 166.

4 Ibid., p. 167; Butler, *The Book of Fines*, p. xxiii.

5 Randall Moffett, 'The Military Organization of Southampton in the Late Medieval Period, 1300–1500' (unpublished doctoral thesis, University of Southampton, 2009), p. 192.

to attack, as was demonstrated on 5 October 1388, when a French fleet attacked and burnt the town.[6] Later, in 1457, the inhabitants were successful in repelling another French raid, in part due to the strength of their defences and artillery.[7] The town was also embroiled in the factional conflict of the Wars of the Roses, as can be seen from a commission of array granted to the aldermen of the town on 28 April 1460, to resist the Yorkist lords and to 'fortify the walls and make defences called "loupes" thereon, and to appoint watches, scouts and keepers of the gates day and night'.[8] The inhabitants also participated in the military operations of the crown, such as in 1470, when soldiers were sent for the king's expedition to the north.[9] Southampton was therefore a prosperous town for much of this period, with well-established rights of self-governance and one which regularly took part in military activities for its own defence and to aid the ruling monarch. Therefore, the aldermen had the means to invest in the latest types of artillery fortifications and firearms to ensure that they were adequately protected.

The principal sources used for this case study are the steward books of Southampton, which detail the income and expenditure of these officials. The steward was an important town official, whose role incorporated providing legal advice to the Mayor's Court, interpreting town charters and acting as treasurer for the town.[10] The survival rate of these books varies greatly by decade, with a relatively high proportion of extant books for the 1470s and 1480s (40 per cent and 50 per cent), whereas the 1440s, the 1450s and 1490s only have a 20 per cent survival rate.[11] This relative abundance of evidence allows a unique insight into the development of urban artillery, although the gaps mean that a complete assessment of Southampton's experience of firearms cannot be established. This chapter uses the extensive surviving evidence for Southampton to examine the development of the technology in the town.

NUMBERS OF GUNS AND DEVELOPMENT OVER TIME

In the first half of the fourteenth century, traditional weapons were used for the defence of the town, with its arsenal possessing twelve springalds and two large

6 Jonathan Sumption, *The Hundred Years War: Volume I, Trial by Battle* (London: Faber, 1990), p. 248.

7 J. Silvester Davies, *A History of Southampton: Partly from the Ms. Of Dr. Speed in the Southampton Archives* (Southampton: Gilbert, 1883), p. 473.

8 *CPR 1452–1461*, p. 602.

9 Davies, *A History of Southampton*, pp. 23–4; SRO SC 5/1/12, ff. 13r–13v.

10 Harry W. Gidden, ed., *The Book of Remembrance of Southampton*, volume 1 (Southampton: Cox & Sharland, 1927), p. xviii.

11 SRO, SC 5/1/1–27B.

mangonels in 1353.[12] Later in the century, the aldermen also began to acquire gunpowder weapons; 5s. 8d. was paid for a gun in 1382.[13] The next available evidence for firearms dates to an inventory of 1434, which shows that five guns were located in a storehouse, consisting of one large gun, two smaller guns and two small pellet guns, together with ammunition and gunpowder.[14] The number of guns had expanded by 1450, in large part due to equipment transferred by Sir Thomas Kyriell, the commander of an expeditionary force, prior to his departure for Normandy. Repairs were subsequently carried out to the donated equipment; eight gun chambers were mended and a cart and lead shot were provided for the organ guns.[15] The accounts for this year also record that six guns stored at God's House Tower were moved to Harry Gunner's house. This edifice appears to have been used as a workshop for creating gun chambers, while a further two guns were kept at the Bargate.[16] Coastal defence was clearly of sufficient importance for Sir Thomas Kryrell to be willing to give some of the guns used in his expedition to the town, which were intended for Normandy in 1450, despite the desperate need for artillery in the duchy.[17] Defensive preparations intensified in 1457 when the town was threatened by a French fleet and numerous expenses were incurred on the deployment and repair of guns. Labourers were hired to transport ordnance from the stores to the walls and bulwark. In addition, gunners were tasked with the making of gunpowder and a smith was employed to refurbish the iron guns.[18]

The aldermen of Southampton clearly appreciated the defensive value of firearms by the beginnings of the 1460s, as can be seen from a letter they addressed to the 'Lords of the Council' in 1460, which claimed that they possessed twenty-five guns but that sixty were necessary for the safety of the town.[19] It appears that their efforts to increase their stockpile of weaponry was successful, as the 1467–8 inventory lists a total of thirty-seven guns, including a large gun called *Thomas with*

12 Randall Moffett, 'Military Equipment in the Town of Southampton during the Fourteenth and Fifteenth Centuries', *Journal of Medieval Military History*, 9 (2011), 167–99 (p. 171).

13 James E. Thorold Rogers, *A History of Agriculture and Prices in England from the Year after the Oxford Parliament, (1259) to the Commencement of the Continental War, (1793)*, volume 2 (Oxford: Clarendon Press, 1866–1902), p. 59.

14 Harry W. Gidden, ed., *The Steward's Books of Southampton, from 1428*, volume 1 (Southampton: Cox & Sharland, 1935), pp. 90–3.

15 Note that the exact number of guns given to the town is unspecified.

16 SRO, SC 5/1/7, ff. 16r, 17v, 19v.

17 Anne Curry, 'Guns and Goddams: Was there a Military Revolution in Lancastrian Normandy, 1415–50?', *Journal of Medieval Military History*, 8 (2010), 171–88.

18 SRO, SC 5/1/8, ff. 12r–14v.

19 R. C. Anderson, ed., *Letters from the Fifteenth and Sixteenth Centuries: From the Archives of Southampton* (Southampton: Cox & Sharland, 1921), p. 21.

FIGURE 23 Types of guns in Southampton inventory of 1467–8.

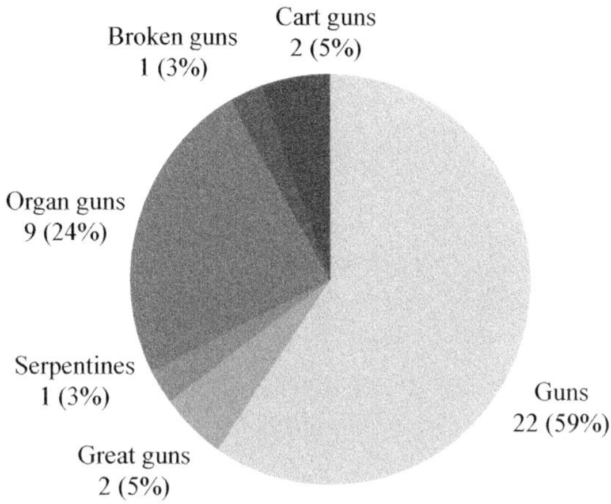

Source: SRO, SC 5/1/10.

the Beard, a serpentine, two cart guns and an organ gun (see Figure 23).[20] Two years later, payments were made to maintain the town's weaponry; repairs were carried out to the ordnance, tampons acquired for the organ guns, and a gun purchased for £1 6s.[21] The following decade saw further activity related to the town's artillery, almost certainly due to hostilities with France in this period. In 1472–3, fear of attack appears to have prompted the deployment of guns to the Quay and the Watergate.[22] Two years later, numerous payments were made for the repair of the ordnance, including *Thomas with the Beard* and the organ guns, some of which were damaged whilst being moved by porters to defensive positions in the town.[23] A reduced level of activity occurred in the 1480s, although repairs were carried out to gun chambers, gunpowder was purchased, and guns were deployed in 1482–3.[24] In the following year, guns were moved from the Bargate to the Watergate and other guns were also transported from the latter to a smith for repairs.[25] By the end of the fifteenth century, however, the town's stockpile of weaponry had decreased dramatically, with only seventeen guns, along with an unspecified number of

20 SRO, SC 5/1/11, ff. 17v–18r.

21 SRO, SC 5/1/12, ff. 12v–13r, 14v, 17r–17v.

22 SRO, SC 5/1/14, ff. 9r, 31v.

23 SRO, SC 5/1/15, ff. 15r, 24r–24v, 26v–27r.

24 SRO, SC 5/1/18, ff. 14v, 15v.

25 SRO, SC 5/1/19, f. 46r.

handguns, recorded in a document of 1498–9.[26] Therefore, the number of guns possessed by the Corporation of Southampton changed significantly over the course of the period, with the 1450s and 1460s marked by a dramatic increase in numbers, due in part at least to royal support.

DEPLOYMENT AND LOCATION OF GUNS

Most of the guns of the town were kept in a centralised storage place described variably as the Mill House, Store House, or Town's Store. It is possible that these refer to different buildings, but only one of these names is used in each of the yearly books. In 1469–70, labourers were paid to fetch guns out of the Mill House, whereas in 1474–5 guns were moved from the Store House to the Quay.[27] In the inventory of 1434, five guns were listed as being stored there, made up of one large gun, two smaller guns and two pellet guns, whereas by 1467–8 the items in the inventory consisted of two cart guns, one organ gun, one large gun, one trestle gun and two guns in the 'windows of the same house', along with gun chambers and gunpowder.[28] Ordnance was often moved to and from the Store House to be deployed for the defence of the town or for repairs, as can be seen from entries in 1474–5, which record that guns were moved from the Quay to the Long House, then to the house of Adrian, a smith, back to the Quay again and then finally to the Store House.[29] The Store House was also used for storing gunpowder, gunstones and other equipment, with the inventory of 1434 listing gunstones for the guns 'without number', two barrels of powder and an iron rod for the said guns.[30] There are also a limited number of references to guns being located in other areas, such as the Long House, to which guns were temporarily moved in 1474–5.[31]

The two main gates of the town, the Bargate in the north and the Watergate to the south, were often furnished with artillery for their defence. In 1449–50 there were at least two guns at the Bargate, whereas by the time of the inventory of 1467–8 this number had increased to three.[32] Guns were also, on occasion, transferred between these two locations; such as in 1483–4 when guns were moved from the Bargate to the Watergate.[33] The latter contained one large gun in the

26 Butler, *The Book of Fines*, p. 2.

27 SRO, SC 5/1/12, f. 13r; SC 5/1/15 f. 26v.

28 Gidden, *The Steward Books of Southampton*, volume 1, pp. 90–3; SRO, SC 5/1/11 ff. 17v–18r.

29 SRO, SC 5/1/15, f. 26v.

30 Gidden, *The Steward Books of Southampton*, volume 1, pp. 90–3.

31 SRO, SC 5/1/15, f. 26v.

32 SRO, SC 5/1/11, ff. 17v–18r; SC 5/1/7, f. 16v.

33 SRO, SC 5/1/19, f. 46v.

inventory of 1467–8, and there are numerous entries relating to guns deployed there, such as in 1472–3 and 1482–3.[34] The importance of this gate can be seen from subsequent evidence from later in the sixteenth century, for provision was made for its defence in 1549, when it was ordered 'Fryst, to fortyfye the Watergate wth baskets and to be furnished with the best ordenance, and a covergate wt sume defence for the tower there'.[35] By contrast, the only other evidence for artillery being deployed to other gates occurs in 1456–7, when a reference was made to a gun that was located at God's House Gate.[36] The predominance of guns at the Bargate and Watergate appears to be a reflection of their importance as the chief entry points into the town from land and sea respectively.

God's House Tower was a vital defensive structure in Southampton.[37] At least six guns were kept in the tower in 1450, and the inventory of 1467–8 records four guns, together with gun chambers, gunpowder and ammunition.[38] This suggests that the tower served as an additional storage area for the Store House, which had its own supply of gunpowder. The inventory of 1467–8 also lists a number of other towers which housed guns: namely, one gun in each of the two towers between Bargate East and St Denys Tower (Polymond), another gun in St Denys Tower, two guns in the tower next to Bargate westwards, one gun in Pilgrims Pyt, one gun in the tower next to the playhouse, two guns in the tower next to God's House Tower to the east, two guns in Gebon Cormonger's Tower, and two guns in Mechell Luks Tower.[39] This layout indicates that guns were considered best deployed at the northern and southern defences of the town. The Quay was also a key location for the deployment of ordnance. A large gun was moved there by a labourer in 1456–7, while other guns were deployed there later in 1474–5 and 1482–3.[40] Significantly, the larger guns were often placed on the Quay, such as *Thomas with the Beard* in 1474–5, which indicates that they were considered to be useful in repelling sea-borne attacks or as a visual deterrent to potential attackers.[41] By contrast, there are few specific references to guns being deployed to the town's bulwarks, as in 1456–7,

34 SRO, SC 5/1/11, ff. 17v–18r; SC 5/1/14, f. 9r; SC 5/1/18, f. 15v.

35 Harry W. Gidden, ed., *The Book of Remembrance of Southampton*, volume 2 (Southampton: Cox & Sharland, 1928), p. 21.

36 SRO, SC 5/1/8, f. 17v.

37 *CPR 1416–1422*, p. 109. See Plate 4.

38 Gidden, *The Steward Books of Southampton*, volume 1, p. 53; SRO, SC 5/1/7, f. 17v; SC 5/1/11, ff. 17v–18r.

39 Ibid.

40 SRO, SC 5/1/8, f. 12v; SC 5/1/15, f. 26v; R. C. Anderson, ed., *The Assize of Bread Book, 1477–1517, Publications of the Southampton Record Society* (Southampton: Cox & Sharland, 1923), p. 41.

41 SRO, SC 5/1/15, f. 26v.

although the bulwarks' specialised function as artillery fortifications suggests that the surviving records may underplay their role.[42]

It appears that the guns of the town were deployed and moved about in a determined fashion by the civic authorities. The Store House was used for storing guns and other equipment, with God's House Tower also fulfilling a similar function. The focus of the defence appears to have been on the northern side (The Bargate), and on the southern side (Watergate, God's House Tower and the Quay). These were also the main entry points into the town from land and sea, for both travellers and attackers. Some idea of the procedure for moving guns from storage into operational use can be seen from the records for the year 1450. On 29 January, six guns were transported from the West Quay to God's House Tower, then on 8 May they were carried to Harry Gunner's House, Harry was then paid on 8 July for the making of seventeen new gun chambers and for the repair of eight existing gun chambers.[43] This suggests that guns were deployed from God's House Tower in periods of crisis and were then subsequently moved elsewhere for repairs.

ARTILLERY FORTIFICATIONS

Southampton was first fully encircled by masonry defences in the fourteenth century, as a result of a destructive French raid in 1338.[44] In the second half of the fourteenth century, between 1360 and 1380, gunports for small guns were inserted into the walls of blocked up merchant houses by the West Quay.[45] This would certainly indicate that the town had guns prior to its earliest recorded reference in 1382, possibly as a result of the stationing of a garrison in Southampton in 1377, under the command of Sir John Arundel, which included an unspecified number of gunners.[46] Gunports are rarely mentioned in the accounts, but a mason was paid to make a hole in the wall at the Bargate for a gun in 1449–50, and payments were made for the mending of 'loop holes' in 1456–7 as well as in 1467–8.[47]

A major addition to the town's defences was made in the early fifteenth century with the creation of God's House Tower. It is unclear when construction work started on the tower, but the structure was built sometime before 1417, when £100

42 SRO, SC 5/1/8, f. 12r.

43 SRO, SC 5/1/7, ff. 17r–17v.

44 Moffett, 'Military Equipment in the Town of Southampton during the Fourteenth and Fifteenth Centuries', p. 167.

45 See Plate 3. John R. Kenyon, 'Early Artillery Fortifications in England and Wales: A Preliminary Survey and Reappraisal', *Archaeological Journal*, 138 (1981), 205–40 (p. 210).

46 TNA, E 403/465, m. 6.

47 SRO, SC 5/1/7, f. 16v; SC 5/1/8, f. 12v; SC 5/1/11, f. 12r.

was granted by Henry V from the customs duties for its repair.[48] This building still survives, located on the south-eastern corner of the town next to the sea, and includes large gunports and a roof platform for ordnance.[49] As previously noted, God's House Tower was an important location for guns by the time of the 1467–8 inventory and was therefore likely to have been the focus of the town's defence. Catchcold Tower, first mentioned in 1438, was also constructed in the first half of the fifteenth century.[50] This building is larger than the other towers in the circuit of the tower walls and includes three gunports; however, no documentary evidence survives for the deployment of guns in the tower itself.

The aldermen of Southampton continued to invest in their fortifications throughout the period. The work which was undertaken included the construction of bulwarks in the second half of the fifteenth century. The earliest mention of a bulwark occurs in 1457, when payments were made for work 'about the bolewark' and other ordnance, including nails and hooks for its door.[51] This building, together with God's House Tower, was clearly important to the defence of the town, as can be seen from a payment for 5d. for '5lbs of candles that were wasted in Godshouse Tower in the bolewark the night the French fleet was there'.[52] A note attached to the letter addressed to the 'Lords of the Council', circa 1460, also makes reference to a non-extant document regarding the 'grev(ou)s costs done uppon the walles of frames of tember abought the boullewerkes'.[53] In 1469–70, a new bulwark was constructed at the Watergate, overseen by William Temple, clerk of the ordnance, who was incorrectly described in the accounts as the master of the king's ordnance.[54] This was carried out by seven men who were paid from two to seven days for carrying out this work. Seven labourers were also hired at the same time for the setting up and filling of the 'pipes' at the bulwark, which were later bound with thirty hoops. On 30 June 1484, three labourers were paid for a day's work upon the bulwark at West Quay and for 'settyng of pylys under the wall by the castell gate', while repairs were later carried out on the timbers of the bulwarks 'without Goddys house gate' on 18 October 1487.[55] By the end of the fifteenth century, the

48 *CPR 1416–1422*, p. 109; Hilary L. Turner, *Town Defences in England and Wales: An Architectural and Documentary Study AD 900–1500* (London: John Baker Ltd, 1971), p. 175; Moffett, *The Military Organization of Southampton in the Late Medieval Period*, p. 43.

49 Moffett, *The Military Organization of Southampton in the Late Medieval Period*, pp. 43–4.

50 Gidden, *The Steward Books of Southampton*, volume 2, p. 83.

51 SRO, SC 5/1/8, f. 12r.

52 Ibid.

53 Anderson, *Letters from the Fifteenth and Sixteenth Centuries*, pp. 21–2.

54 For instance, see Exchequer accounts where he is described as a clerk of the ordnance, such as TNA, E 405/57, m. 1; SRO, SC 5/1/12, ff. 17r–17v..

55 SRO, SC 5/1/19, f. 37v; SC 5/1/22, f. 24r.

town was in possession of a sizeable number of bulwarks, which suggests that they were considered crucial to its defence.

The development of artillery fortifications in Southampton can therefore be seen to fall into three phases. Firstly, in the late fourteenth century the existing defences were modified for the use of guns through the insertion of gunports into the walls. Secondly, in the early fifteenth century specially designed towers, God's House Tower and Catchcold Tower, were constructed to maximise the use of these weapons. Finally, in the mid-fifteenth century multiple bulwarks were constructed at the quays and gates of the town. This would suggest that Southampton's experience of gunpowder weapons was broadly similar to that of other coastal towns in the south-east of England, such as Sandwich and Rye. This can be seen with the insertion of gunports in the late fourteenth century, as at Canterbury in 1380, and the construction of bulwarks in the second half of the fifteenth century, as at Great Yarmouth in 1457.[56] The construction of artillery towers in the first part of the fifteenth century, however, suggests that the development of the technology in Southampton differed from other similar towns in the region. These structures were not unique in England, as is shown by the late fourteenth-century examples of the Cow Tower in Norwich and the Westgate in Canterbury. Nevertheless, the decision to build these structures shows that the burgesses of Southampton continued to adapt their defences to gunpowder weapons at a time when few other towns appeared to do so.

PERSONNEL

The Corporation of Southampton were unusual in employing an official gunner, referred to in the accounts as the 'Town's gunner', from the middle of the fifteenth century.[57] This is a role that changed over time, with the earliest gunners provided with livery and only paid on an ad hoc basis, whereas the gunners of the sixteenth century were paid salaries. The first recorded gunner, Harry Gunner, only received payment for the making and mending of gun chambers, which suggests that he was not considered an official of the Corporation.[58] Unusually, however, a series of payments were made for the repair of his house, which came to the sum of £1 16s. 5d., and consisted of payments to carpenters, masons and roofers.[59] It is possible that the town had an arrangement with Harry to provide accommodation in lieu of payment which has not survived in the documentary record. Certainly some

56 B. H. St J. O'Neil, *Castles and Cannon: A Study of Early Artillery Fortifications in England* (Oxford: Clarendon Press, 1960), pp. 207–12; TNA, E 364/93, rot. F.

57 SRO, SC 5/1/7, f. 17v.

58 Ibid., f. 17v.

59 Ibid., f. 9v.

contracts with gunners in the fifteenth century included accommodation as well as wages, as can be seen by an entry in the accounts of John, Lord Howard, which records such an arrangement with his gunner in 1463.[60] Entries in the steward books suggest that Harry used his house as a workshop, as guns were moved there to be repaired and refurbished.[61] The next gunner mentioned is John Bronne, who was paid 6d. a day for a total of eight and a half days in 1456–7; in the same year a John Gunner from Sandwich was paid 12d. for his labour.[62] This is of interest as Sandwich was itself attacked in 1457 and John Gunner is next mentioned in 1461–2 as in receipt of livery, undoubtedly as the town gunner.[63] He had a long association with the town beginning in 1461–2 until 1482–3, and although he did not receive a salary, again received livery in 1467–8, 1469–70, 1470–1 and 1482–3.[64] In addition to ad hoc payments for gunnery work, he was paid in 1461–2 and 1467–8 for work on the conduit, and in 1469–70 for cleaning the Audit House and for work on the sea-hedge.[65] The latter examples suggest that he was forced to supplement his income through odd jobs, when there was insufficient work for him to be employed as a gunner by the Corporation. Furthermore, the surviving sources do indicate that whilst the town often felt the need to employ a gunner, there were also periods when this was not the case.[66]

Town gunners appear to have been individuals with expertise in different areas, such as masonry techniques in carving gunstones, the ability to create gunpowder, and smithy skills in the repairing of wrought iron guns. The clearest evidence for the intended role of the gunner dates from 1550, where it was specified that 'He shall at (all) tymes be redy to overse the ordering of the townes ordynance and redy to do the commandment of the meyer of the town for the tyme being' yn all things abowt' the said ordynance'.[67] In addition to the town gunner, other gunners were employed on a temporary basis, particularly in times of conflict, such as John Gunner who was hired when Southampton was threatened by a French fleet in 1457.[68]

Gunners were town officials who were also granted livery. The earliest evidence for this comes from a payment of 1461–2 for three yards of cloth for 7s. for John

60 Anne Crawford, ed., *The Household Books of John Howard, Duke of Norfolk, 1462–1471, 1481–1483* (Stroud: Sutton for Richard III & Yorkist History Trust, 1992), I, 225.

61 SRO, SC 5/1/7, f. 17v.

62 SRO, SC 5/1/8, ff. 12r–12v.

63 SRO, SC 5/1/10, f. 13r.

64 SRO, SC 5/1/12, f. 21v; SC 5/1/11, f. 1r; SC 5/1/13, f. 33v; SC 5/1/18, f. 26r.

65 SRO, SC 5/1/10, f.13r; SC 5/1/11, f. 1r; SC 5/1/12, f. 18r.

66 For instance there is no mention of a town gunner in 1474–5, despite numerous references relating to ordnance that year, SRO, SC 5/1/15.

67 Gidden, *The Book of Remembrance of Southampton*, volume 2, pp. 30–1.

68 SRO, SC 5/1/8, f. 12v.

Gunner.[69] This also gives some insight into the status of the gunner relative to other officials of the corporation, as this was the same amount of material given to the crane man and one yard less than was given to each of the sergeants.[70] Later in 1467–8, John Gunner was given three yards for his gown, priced at 2s. 4d. per yard, whereas the four sergeants were given eighteen violet yards, priced at 3s. 4d. per yard.[71] In 1469–70 he was issued with two and half yards of cloth of Montivilliers, price 3s./yard, whereas the four sergeants each received four yards, priced at 3s. 4d./yard.[72] In 1470–1 he received two and a half yards of broad cloth of 'rone russet' for his livery gown, priced at 3s./yard, the sergeants received four and a half yards of violet broad cloth, priced 3s. 4d./yard.[73] The gunners also used the services of assistants to aid them in their work. There is little evidence for the activities of these individuals, or even their names, but they seem to have been used to assist the gunner in his duties. For instance, two men were paid 5d. in 1456–7 to assist John Bronne in making gunpowder.[74] Other workers were also occasionally hired for work on the ordnance, such as the labourers employed to charge guns in 1474–5.[75]

Therefore, the early gunners were transient in nature. The first two gunners were employed in the particularly tense years of 1449–50 and 1456–7, possibly only on a temporary basis. John Gunner had a long relationship with the town, being in receipt of livery and ad hoc payments for work; however, it was not until the sixteenth century that the gunners of the town received an official salary. This appears to have reflected a change in attitudes towards the management of the ordnance, which was managed on a more professional rather than an ad hoc basis. The Books of Remembrance show that by the sixteenth century the employment of gunners was considered crucial for the interests of the town. In 1512, Balfer Lynne was brought before the Mayor and other town officials and swore to serve the town faithfully in return for a salary and livery. In 1527 Roger Cobley was employed on a similar basis, with the provision that he was 'to be at all tymes redy to serve the towne yn the said rome, and not to absent hym self yn warre or peax for any other profit wtowt licens of Master Meyer'.[76] In a 'provision for the good order and surety of the town' in 1517, it was specified that the town's ordnance should be in readiness and gunners should be appointed.[77] At least some of the gunners, such as John

69 SRO, SC 5/1/10, f. 19r.

70 Ibid.

71 SRO, SC 5/1/11, f. 19r.

72 SRO, SC 5/1/12, f. 21v.

73 SRO, SC 5/1/13, f. 33v.

74 SRO, SC 5/1/8, f. 11v.

75 SRO, SC 5/1/15, f. 26v.

76 A. L. Merson, ed., *The Third Book of Remembrance of Southampton, 1514–1602, volume 1* (Southampton: University of Southampton, 1952), pp. 41–2.

77 Ibid., p. 53.

Gunner, also appear to have started their careers on a temporary basis before later becoming the town's permanent gunner.

ROYAL POLICY

The earliest evidence for guns at Southampton dates from November 1377, when Sir John Arundel was paid for the wages of a garrison at the town which included gunners, although it is unknown how many guns were used in the defence of the town at this time. [78] The crown also maintained its own stock of guns in the royal castle at Southampton. The earliest reference to this appears to date from 5 July 1386 when a carter was paid to transport guns and other artillery for the defence of the castle and town of Southampton.[79] Seventeen days later, a chaplain, Thomas Tredyngton, was appointed to serve the king in his 'new tower of Southampton' to celebrate divine services and to provide for its defence, including the guns of the tower.[80] This was a new tower constructed at 'Old Castle Hill' between 1378 and 1388, which was to include two gates and 'a mantlet with a barbican round the tower'.[81] Chaplains continued to be appointed to this position throughout the fifteenth century; limited evidence exists for the numbers of guns in their possession, but in 1386 the arsenal consisted of three guns with 200lbs of gunpowder, whereas by 1437–42 it was made up of two large guns.[82]

It also appears that it was royal policy to support the development of guns in Southampton. Although no grants were made for purchases of ordnance, some were given for repairs at God's House Tower in 1417 and for expenditure on the walls of Southampton in 1469 and 1478.[83] In 1460 the aldermen of Southampton were commanded to work on their fortifications and to provide loops, presumably for the use of guns.[84] This command might have prompted the sending of a letter to the 'Lords of the Council' by the aldermen, which stressed the weakness of the walls to gunfire, and the lack of ordnance.[85] It may have been a response to this which led to the great expansion of guns in this period, either due to the efforts of the Corporation or to government assistance. Edward IV was interested in the defence

78 TNA, E 403/465, m. 6.

79 TNA, E 403/512, m. 13.

80 *CPR 1385–1389*, p. 196.

81 O'Neil, *Castles and Cannon*, pp. 10–11; *CPR 1452–1461*, p. 20.

82 Moffett, 'Military Equipment in the Town of Southampton during the Fourteenth and Fifteenth Centuries', pp. 177–8; Tout, 'Firearms in England in the Fourteenth Century', p. 699; TNA, E 364/76, rot. H.

83 *CPR 1416–1422*, p. 109; *CPR 1467–1477*, pp. 154–5; *CPR 1476–1485*, p. 76.

84 *CPR 1452–1461*, p. 602.

85 Anderson, *Letters from the Fifteenth and Sixteenth Centuries*, p. 22.

of the town, as can be seen from a reference to a payment to William Temple, clerk of the ordnance, for supervising the construction of bulwarks in 1470. These works were carried out by the king's command as can be seen from a royal writ of the same year which refers to 'the charges that they by oure comaundement must bere in stokkyng of gunnes and makyng of bulwerks for the defense of oure saide towne'.[86] Further royal investment occurred in the reign of Henry VII, with the founding of an arsenal and royal dockyard in Southampton and Portsmouth respectively.[87] In the war of 1492 ordnance was shipped from Southampton to France, with royal guns and gunpowder stored in the castle.[88] There is also some evidence of an ordnance industry in the town itself, which supplied guns to the crown, particularly for royal ships. On 4 June 1486, a Philip Lokier of Southampton, smith, was paid to deliver eight serpentines to the king's ship *The Mary of the Tower*, for a total of £8.[89] The purchase of twenty guns and two serpentines in 1484 in Southampton by Thomas Rogers, clerk of the king's ships, seems to corroborate this, although they may have been imports from abroad.[90]

CONCLUSION

The financial records demonstrate the crucial importance of gunpowder weapons to the defence of the town in the fifteenth century. Initially the possession of these guns was largely carried out under the auspices of the crown, with guns and gunners sent to the town and its castle in the crisis years of 1377 and 1386. From the early fifteenth century onwards, however, it appears that the focus shifted to the burgesses making use of their own artillery for their defence. Then from 1434–68 there was a rapid expansion in the number of guns, yet the end of the century saw a period of decline. The initial expansion appears to have been motivated by the turbulence of the middle part of the fifteenth century, when the town was threatened by both internal and external enemies. It is however difficult to account for the reduction by the end of the period, as the threat of invasion remained in the 1490s.

The development of gunpowder weapons in Southampton is broadly similar to that of other towns in that it appears to have been motivated by a response to

86 SRO, SC 5/1/12, f. 17r; TNA, E 159/247, brevia directa baronibus, Trinity, rot. 12.

87 Silvester, *A History of Southampton*, p. 476.

88 TNA, E 36/285, f. 5r.

89 M. Oppenheim, ed., *Naval Accounts and Inventories of the Reign of Henry VII, 1485–8 and 1495–7* (London: Naval Records Society, 1896), pp. 19–20.

90 Rosemary Horrox and P. W. Hammond, eds., *British Library Harleian Manuscript 433. Vol. 2, Second Register of Richard III* (Gloucester: A. Sutton for the Richard III Society, 1980), p. 112.

external threats. Yet the town is unusual in having evidence of expenditure on guns and artillery fortifications almost continuously from the late fourteenth century to the end of the period. This suggests that despite similarities in the ways in which many of the coastal towns of southern England adopted and made use of these weapons, there was a degree of variety. Southampton's prominent use of ordnance could be due to the relatively high survival rate of steward books. The evidence indicates, however, that the Corporation of Southampton was exceptional in its commitment to firearms. This can be seen in particular by the intermittent employment of a town gunner from the mid-fifteenth century onwards, as no evidence exists for the official employment of gunners by other towns before the sixteenth century, although York and Sandwich did have keepers of artillery. The example of Southampton shows that English towns did not adopt gunpowder weapon technology in a uniform fashion.

Part III

Analysis

8

Analysis of Guns

THE ADOPTION OF GUNPOWDER weapons was one of the major developments of the Middle Ages. Establishing how this technology was adopted and adapted by medieval artificers is more difficult to determine, however, due to a lack of detail in the extant financial sources. Yet the preceding chapters have demonstrated that extensive evidence survives from other sources for English firearms in the late fourteenth to the fifteenth centuries. The information extracted from these sources can be used to answer key questions about the construction, maintenance, operation and movement of these guns, as well as the types of gunpowder and ammunition used to fire these weapons. This chapter explores these topics in detail.

BRONZE GUN-MAKING IN ENGLAND

The earliest record of firearms in English accounts dates from 1345, when an unspecified number of guns were repaired for the royal expedition to France, but the provenance of these weapons is unknown.[1] Similarly, the process by which gun-making techniques were imported from the Continent is undocumented, although this had occurred by at least 1353, when four guns made out of a copper-alloy were produced for the privy wardrobe by a brasier, William de Algate.[2] A variety of terms were used to describe guns made of copper-alloys, including copper (*cupra*), brass (*aenum/eneus*), laton (*latone*) and metal (*metallo*).[3] This production

1 T. F. Tout, 'Firearms in England in the Fourteenth Century', *The English Historical Review*, 26 (1911), 666–702 (p. 688).

2 Note that the modern term bronze is used in this book to refer to copper-alloy guns in general. For a discussion of this see Claude Blair and John Blair, 'Copper Alloys', in *English Medieval Industries*, ed. John Blair and N. Ramsay (London: Hambledon Press, 1991), p. 81; Tout, 'Firearms in England in the Fourteenth Century', p. 691.

3 The latter was sometimes used to distinguish iron and copper-alloy guns. For instance, the accounts of John Hampton for 1430–2 include the entry 'xvj canonis de ferro vocatur fowellers ... iiij canonis de metallo vocatur fowelers', TNA, E 364/69, rot. Q

method involved heating copper, along with another metal, almost always tin, at a sufficiently high temperature until it became liquid, after which it was then poured into a mould, usually of metal or clay.[4] A range of different words were used to describe craftsmen who worked with copper-alloys, such as potters, latteners and founders, although the earliest recorded gun-makers appear to have been London brasiers, such as John Brassier of Cornhill, who made one gun for the privy wardrobe in 1361.[5] As has been suggested by previous scholars, these professionals were pioneers in gun production as the techniques they employed for casting pots and cauldrons could easily be adapted for making early firearms.[6] An example of this can be seen in the case of the gun depicted in the manuscript presented by Walter de Milemete to Edward III in 1327, which takes the form of a cauldron placed on its side on a wooden trestle.[7] By the late fourteenth century, however, the vast majority of bronze guns were instead manufactured by founders, who were primarily employed in the complex task of casting bells for religious institutions. It is possible that this change was prompted by technological advances in gun-making, which required more specialist skills, but a lack of evidence means that this claim cannot be substantiated.[8]

The most notable known English founder of the late fourteenth and early fifteenth century involved in the production of guns for the crown was William Woodward, a citizen of London and resident of St Botolph's Aldgate. Woodward can be first detected in royal service on 1 May 1385 when he supplied twelve guns for the defence of Dover Castle, while a further sixty guns had been manufactured for Calais by 24 February 1386.[9] He was later rewarded, on 22 December 1395, with a grant for life of the keepership of the great clock within the Palace of Westminster, with a daily wage of 6d. in return for ensuring its maintenance.[10] Woodward continued to be employed by the crown during the reign of Henry IV, being appointed on 26 April 1401, along with Gerard Sprong, to take metal and workers for the production of guns in London and elsewhere.[11] He evidently prospered in royal service, as by 24 November 1406 he was described as a king's esquire when a

dorse; Tout, 'Firearms in England in the Fourteenth Century', pp. 691–7; E 364/49, rot. C; E 101/51/27; E 405/69, m. 3d; E 405/566, f. 61r; E 101/198/13, ff. 3r–13r; E 36/8, ff. 1r–7v.

4 Blair and Blair, 'Copper Alloys', pp. 83–7.

5 Ibid., p. 93; Tout, 'Firearms in England in the Fourteenth Century', pp. 691–2.

6 Blair and Blair, 'Copper Alloys', p. 99.

7 Hugh Pollard, *Pollard's History of Firearms* (London: Country Life Books, 1983), p. 28.

8 For instance the earliest recorded weight for an English gun dates to the mid-1380s. Tout, 'Firearms in England in the Fourteenth Century', pp. 697–8.

9 TNA, E 403/508, m. 8; E 403/510, m. 27.

10 Note that it appears that Woodward did not receive this grant, however, as below the entry is the words 'vacated because nothing was done herein', *CPR 1391–1396*, p. 648.

11 *CPR 1399–1401*, p. 480.

commission of mainprise was ordered into his management of a manor in Essex during the minority of its owner, Thomas Enfeld, a relative of his wife Agnes.[12] The management of this estate was removed from Woodward's custody on 20 May 1410, but his skills continued to be valued by Henry V, who granted him an annual gift of livery vestments every Christmas on 13 December 1415 'for his good service'.[13] Five days earlier, the king had given a commission to him and Gerard Sprong to take metal pots, bowls, kettles and other vessels from the royal kitchens for the construction of guns, together with instructions to take timber, saltpetre, coal and workers for the same purpose from London and elsewhere.[14] Woodward continued to play an important role in supplying guns, gunstones and gunpowder for expeditionary armies throughout the reign of Henry V, receiving over £250 from the Exchequer from 17 January 1415 until 3 October 1420.[15] By the time that his will was drawn up on 7 October 1421 he had clearly become a wealthy man, as its terms specified bequests of over £300 in cash, including £67 to thirteen servants who had worked in his foundry.[16] Woodward's relationship with the crown was exceptional, although other founders were occasionally contracted to supply ordnance, such as Thomas Coston and Robert Warner, who provided sixty-four guns for the 1428 expedition to France.[17]

From the 1420s onwards the majority of guns produced in England were made out of wrought iron, yet bronze continued to be used, albeit on a much reduced scale from the late fourteenth century.[18] Specialist foundries capable of manufacturing ordnance were located across the country, with notable centres, such as Bristol and Bury St Edmunds, operating in the second half of the fifteenth century.[19] Nevertheless, the majority of bronze guns purchased on behalf of the crown during the reign of Edward IV were sourced from Antwerp and Brussels in Brabant.[20] The preference shown for ordnance manufactured in the Low Countries suggests that Continental foundries may have had a technological advantage at

12 *CPR 1405–1408*, p. 275.

13 *CPR 1408–1413*, p. 231; *CPR 1413–1416*, p. 270.

14 *CPR 1413–1416*, p. 292.

15 TNA, E 403/619, m. 11; E 403/624, m. 4; E 403/636, mm. 1, 4; E 403/646, m. 1.

16 TNA, PROB 11/2B, f. 400.

17 Dan Spencer, 'The Provision of Artillery for the 1428 Expedition to France', *Journal of Medieval Military History*, 13 (2015), 179–92 (p. 184).

18 See Figure 20.

19 Mary Dormer, ed., *The Coventry Leet Book, or, Mayor's Register, Containing the Records of the City Court Leet or View of Frankpledge, A.D. 1420–1555, with Divers Other Matters* (Oxford: Oxford University Press, 1907–13), p. 260; E. W. W. Veale, ed., *The Great Red Book of Bristol*, volume 1 (Bristol: Bristol Record Society, 1931), p. 131; Sally Badham and John Blatchly, 'The Bellfounder's Indent at Bury St. Edmunds', *Proceedings of the Suffolk Institute of Archaeology and History*, 36 (1988), 288–97 (p. 295).

20 For instance, see TNA, E 101/55/4, ff. 12r–12v.

this time, although some of these weapons were produced in England for the expedition to Scotland in 1481 by the gunfounder William Nele.[21] A significant change took place during the reign of Henry VII with bronze once again becoming the preferred material for gun-making.[22] This technological development appears to have originated in France with the creation of new types of guns, although these techniques were soon transferred to England by experts from Brittany who were active in the production of artillery in the Tower of London in the late 1480s.[23] These changes undoubtedly helped to stimulate the domestic industry with two founders, William Newport and William Forse, later supplying twenty-three falcons and one serpentine for the expedition to Scotland in 1497.[24]

WROUGHT IRON GUN-MAKING IN ENGLAND

The earliest evidence for the production of iron guns in England occurs in 1372, when a Stephen Smith was paid for supplying two guns to the privy wardrobe, although they were made in relatively small numbers before the fifteenth century.[25] These weapons were manufactured using wrought iron, an alloy that is known for its ductile and malleable properties, which means that it can be welded when heated to a sufficient temperature. A type of furnace known as a bloomery was used to extract workable iron by smelting the ore, which involved placing layers of charcoal in the bloomer and then firing, with air circulated by means of bellows.[26] The wrought iron could then be used to forge guns by a process of hammer welding strips of metal together on an anvil, with the metal heated to approximately 1,100 C. The weapons were constructed in two parts, consisting of barrels and powder chambers, which were assembled by binding long bars, known as staves, with hoops and bands. The latter two objects were made by means of wrapping iron around an object called a mandrel and then hammer welding the ends together.[27] These techniques appear to have been adopted with relative ease by smiths throughout England, as many of the methods were familiar to them. This explains why iron became the preferred metal for making guns for much of the fifteenth century, due to the ready availability of smiths, who were present in sizeable numbers across the

21 TNA, E 405/566, ff. 61–2.

22 TNA, E 36/15, f. 8r; E 36/8, ff. 1r–7r, 76r; E 36/124, ff. 37v, 48r, 77r, 82v.

23 TNA, E 36/124, f. 39r.

24 TNA, E 36/8, ff. 58r, 66v, 86r.

25 Tout, 'Firearms in England in the Fourteenth Century', pp. 693–4.

26 Jane Geddes, 'Iron', in *English Medieval Industries*, ed. John Blair and N. Ramsay (London: Hambledon Press, 1991), pp. 168–74.

27 Robert D. Smith, 'The Technology of Wrought-Iron Artillery', *Royal Armouries Yearbook*, 5 (2000), 68–79 (pp. 68–71).

country, the low cost of iron compared to bronze and an increase in the size of guns in the reign of Henry IV.[28]

Detailed information on how iron guns were manufactured is available from the accounts of William Hickling, which lists the payments incurred in the construction of thirty-five guns between 1453 and 1456. These works were carried out at a rented storehouse at Mile End near Whitechapel, by a team of at least five smiths who forged the guns, in addition to a further three men who were paid to operate four forges.[29] The tools used by this workforce included five mandrels, four anvils, three blocks of wood for anvils, four forges, four pairs of bellows, twenty-eight mallets and four shovels; with further payments made for iron, steel and charcoal.[30] Further insights into the process of gun-making can be gleaned from the accounts of John Wode, master of the ordnance, which records works carried out at the Tower of London between December 1471 and April 1472. A bastard gun called *The Long Edward* was forged from 12 January to 16 April 1472 by a team of ten smiths, directed by a Master Peter Poundedunyng, using a ton of iron (2,500lbs), each man being paid for 73 days' work. The items purchased for the operation of the forge and hearth included twelve quarters of charcoal, two gallons of oil for the bellows, one ton of water, four carts of sand, one crane, together with brooms, holders, pots, hammers, hand spokes and ropes. During the same period, another nine smiths were supervised by Poundedunyng in the creation of gun chambers for fowlers and serpentines, in addition to five labourers who were employed for 40 to 130 days on a variety of tasks, including the repair and cleaning of the forge.[31] These examples show that the process of manufacturing iron guns was a complex and time-consuming task, which required a variety of tools and the services of a number of skilled workers.

TRANSPORTATION

The deployment of artillery on campaigns meant that significant resources needed to be allocated to the transportation of guns and their equipment, with animals used to pull carts overseen by carters. In the early fifteenth century the preferred method was to use teams of oxen and horses to move large carts filled with ordnance. The

[28] For the low cost of iron see Appendix B, whereas for the increase in the size of guns see Appendix A.

[29] Three smiths are listed by name in addition to 'diverse' others.

[30] Dan Spencer, 'The Lancastrian Armament Programme of the 1450s and the Development of Field Guns', *The Ricardian*, 25 (2015), 61–70 (pp. 63–4); TNA, E 364/90, rot. D.

[31] Dan Spencer, 'The Tower of London and Firearms in the Reign of Edward IV', *Arms & Armour*, 13:2 (2016), 98–110; TNA, E 101/294/19.

guns would then be unloaded and placed on wooden beds or stocks when they had reached their destination. For instance, in 1418, 100 oxen and 320 horses were used to draw twelve large carts, whereas 60 oxen and 24 horses were used to transport four large carts three years later.[32] The adoption of individual carts for guns in the mid-fifteenth century for the deployment of artillery on the battlefield led to a preference for using horses, as they were faster and easier to manoeuvre than oxen.[33] These were generally described as being 'hackneys' in the reign of Edward IV, although 120 'draught' horses were used in the Scottish campaign of 1482, while for Henry VII's expedition to France in 1492 it was specified that 'chariot', 'sumpter' and 'draught' horses should be purveyed for the carriage of the royal artillery.[34] It appears that specific numbers of horses were used to transport particular types of guns, thus three sets of harnesses were allocated for each serpentine and eight for each bombard for the 1475 expedition to France.[35] Later, in 1497, the ordnance left behind at Berwick included 25 sets of harnesses for one large curtow, 19 for each demi-curtow, 6 for each serpentine, 3 for each falcon and 6 for each 'fare' cart.[36] This meant that large quantities of equipment were needed for the transportation of artillery. An inventory of the field train in Calais in 1485 listed 88 carts for moving guns and another 80 for the baggage, in addition to 502 horse harnesses, 305 horse collars, 87 saddles, 19 riding crops and 10,000 horseshoe nails.[37] Substantial amounts of foodstuffs were also required to feed the animals, as can be seen from an order to supply 300 quarters of oats, beans and peas with hay, litter and sufficient pasture for 60 oxen and 24 horses in 1420.[38]

Carters comprised an essential part of ordnance companies; two master carters and ten other carters were hired for the 1430 expedition to France.[39] On the march they were also assisted by other workers, such as cartwrights, wheelwrights and labourers, in the maintenance and transportation of their vehicles.[40] This was necessary as guns could be transported over long distances during the course of military operations. Part of Henry VII's artillery was transported for over 142 miles on land during the Blackheath campaign of 1497.[41] The carts used varied in size with the larger vehicles, such as those used for bombardelles, having four

32 *CPR 1416–1422*, pp. 133–4, 262.

33 For a discussion of this see Chapter 2.

34 *CPR 1461–7*, pp. 36, 183, 303; *CPR 1467–1477*, pp. 163, 338, 398; *CPR 1476–1485*, pp. 250, 387; *CPR 1485–1494*, p. 404; TNA, PSO 1/52, no. 2697.

35 TNA, E 101/55/4, ff. 29v–31v.

36 TNA, E 36/8, f. 108r.

37 TNA, E 364/119/36, rot. B.

38 *CPR 1416–1422*, p. 262.

39 TNA, E 364/69, rot. Q.

40 For an example of this see *CPR 1467–1477*, p. 127.

41 See Appendix I.

wheels, whereas smaller models, such as carts designed for falcons, only had two wheels.[42] Security was sometimes provided for artillery with three armed men assigned to escort a gun from Blackheath to Canterbury in 1471, while watchmen were hired in 1497 to guard the royal guns whilst they were left at Mile End for test firing, at Tower Wharf prior to being loaded on ships and at Hounslow Heath before the battle of Blackheath.[43] Labourers were also employed to move ordnance by hand, particularly in castles and towns: a wooden sled was used to move four guns from Tower Hill to Tower Wharf in 1428.[44] Particularly large guns were lifted by means of cranes referred to as 'gynnes', whilst being moved to and from carts or onto ships.[45] For example, the inventory of ordnance transferred from the Tower of London to Calais in 1475 included 'a gynne of iiij peces called a stradill gynne to carte and uncart with the said grete gonnes'.[46] The main method for transporting guns for expeditions to France and Scotland was by sea, with five ships used to move the artillery train to Normandy in 1428.[47] Specialist equipment and food stuffs were necessary for ensuring the well-being of horses freighted by water, as can be seen from the list of expenses incurred in 1492 for 'horses of the ordnance upon the sea'. This included 450 cart loads of hay, 1,500 quarts of oats, 332 slings, 12 'brigges' and 80 barrels of fresh water.[48]

THE OPERATION, REPAIR AND MAINTENANCE OF GUNS

The effective operation of a gun required that it first be placed on a stable base, such as a wooden trestle, stock or wheeled carriage, by labourers or other workers.[49] Loading procedures differed for breech-loaders and muzzle-loaders. The former used separate powder chambers called gun chambers, which were filled with gunpowder, then inserted into the back of the barrel and sealed in place with a wooden wedge, known as a forelock.[50] The projectile itself was either loaded from the front of the barrel or from the rear before the powder chamber was added.[51] By contrast,

42 TNA, E 36/8, ff. 1r–6r.

43 CCA, CC/F/A2, f. 43r; TNA, E 36/8, ff. 51v, 54v, 65r, 76v.

44 SRO, SC 5/1/8, f. 12v; TNA, E 36/8, f. 78r; Spencer, 'The Provision of Artillery for the 1428 Expedition to France', p. 188.

45 For examples of this at the Tower of London and Calais see TNA, E 101/294/19; E 364/119/36, rot. B.

46 TNA, E 101/55/7.

47 Spencer, 'The Provision of Artillery for the 1428 Expedition to France', p. 183.

48 TNA, E 36/285, f. 5r.

49 For examples of this see TNA, E 101/198/13, f. 13v; E 36/8, f. 1r; E 364/69, rot. Q dorse.

50 For examples of this see TNA, E 364/59, rot. C; SRO, SC 5/1/15, f. 26v.

51 Robert D. Smith and Kelly DeVries, *The Artillery of the Dukes of Burgundy 1363–1477* (Woodbridge: Boydell Press, 2005), p. 258.

charging ladles made out of iron were used to measure and insert gunpowder into the front of muzzle-loaders, which was then pushed to the back of the barrel by staves, with the shot also loaded from the front.[52] Wooden wads known as tampons were employed to separate the gunpowder from the projectile for both types of gun, to prevent the ammunition being damaged during firing.[53] The gun was then secured in place by means of wooden blocks before an iron rod was inserted into the touch hole to ignite the gunpowder.[54] Breech-loaders appear to have had a faster rate of fire than muzzle-loaders, as their powder chamber could be quickly removed by use of a hammer and then be replaced by a new chamber after the barrel was cleaned.[55] Little evidence on this subject survives but cleaning procedures must have been important to ensure that accidental discharges did not occur.

On campaign it appears that gunners were present at sieges and on the battlefield to direct the firing of guns by teams of labourers. For instance the ordnance company for the Blackheath campaign numbered 236 men, including 49 gunners and 122 pioneers.[56] The presence of gunners was not always necessary for the operation of these weapons, however, as can be seen by the employment of James Tynker at Dover in 1479–80, who was paid to fire the town's guns at French ships.[57] It is unclear how men were trained to use ordnance, although an entry from the Paston Letters, which refers to four soldiers hired in 1468, suggests that these were skills that were acquired by experienced professionals. It was said of these men that 'they kan wele schote bothe gonnys and crossebowes and amende and strynge them, and devyse bolwerkys, or any thyngs that scholde be a strenkthe to the place'.[58] Protection was provided to the gun crews during military operations by large wooden screens constructed by carpenters known as mantlets, together with smaller shields called pavises.[59]

Guns and their accessories needed to be maintained and repaired to ensure that they could be used effectively when required. The wear and tear inflicted by the rigours of campaigning, destruction caused by vermin or fire, together with the rusting of iron and rotting of wood caused by water damage, meant that ordnance

52 TNA, E 101/198/13, ff. 42v, 45r, 81r; E 364/119/36, rot. B; E 36/8, ff. 1v, 18r, 91v, 102r.

53 For examples of this see Tout, 'Firearms in England in the Fourteenth Century', pp. 696–700; TNA, E 101/198/13, f. 81r.

54 Tout, 'Firearms in England in the Fourteenth Century', pp. 696–700; TNA, E 364/49, rot. C; E 36/8, f. 91v.

55 C. S. Knighton and D. M. Loades, eds., *The Anthony Roll of Henry VIII's Navy* (Aldershot: Ashgate, 2000), p. 13.

56 TNA, E 36/8, ff. 78v–81r, 82r.

57 BL, MS Add 29616, f. 208r.

58 James Gairdner, ed., *The Paston Letters: A.D. 1422–1509*, volume 4 (London: Chatto & Windus, 1904), p. 306.

59 Frank Taylor and John. S. Roskell, eds., *Gesta Henrici Quinti* (Oxford: Oxford University Press, 2004), p. 37; TNA, E 101/294/19.

periodically had to be refurbished. Contemporaries were certainly aware of these issues, as can be illustrated by a letter sent by Edward IV to Richard Neville, earl of Warwick, captain of the Calais garrison, on 20 August 1465. The king wrote that he had been 'duely enformed that of oure stuffe of artillarie abilements of werr ... some therof be hold and feble', as a consequence of this without 'amending and translatyng they be nen like to serve'. Warwick was therefore instructed to 'suffisauntly repayer amende and translate all the forseid stuffe of artillarie', although it is not possible to determine from the surviving accounts of the treasurer and victualler whether these repairs were carried out.[60] Five years later, Edward wrote to the master of his ordnance, John Wode, as the campaigns of 1470 and 1471 had meant that the royal artillery was 'spoyled and the residue therof broke and brosed in such feldes as we of late had'. Wode was therefore instructed to purchase all 'necessaries in any wise nedefull or containing to our said ordeunce and artelerie as by your discrescon now at this tyme to be repayred or now made shalbe thought expedient for the said defense of our personne and of this our royine'.[61] Civic authorities were also keen to ensure that their valuable guns should be sufficiently repaired and maintained. This can be seen at York in 1484, where it was alleged that the keeper of the artillery, John Craven, had been deficient in carrying out his duty of maintaining the guns and had instead 'suffered them to rust and be lost' and so should lose his fee.[62] In the same year, the decision was taken by the citizens of London that a survey should be carried out of the city's ordnance, following a fire, and that any damaged guns or gunpowder should be repaired or replaced.[63]

Guns constructed out of bronze were stronger than those made of iron, but once broken they could not be repaired and the metal had to be melted down. By contrast, most components of an iron gun could be mended or replaced by a smith or gunner on a forge. The detailed entries relating to expenses incurred on ordnance by some of the towns, reveals that certain items were particularly prone to breakage or deterioration, such as gun chambers, stocks and forelocks. For instance in 1474–5, repairs were carried out to the guns belonging to the burgesses of Southampton, with the works including replacement of two forelocks and a stock for a large gun called *Thomas with the Beard*, as well as mending five chambers of the organ gun that had been broken when moved.[64] A later set of accounts lists the restoration work carried out on a gun at Dover in 1477–8, which included removing the gun from its old stock, repairing its chamber with newly purchased

60 Note that the treasurer and victualler accounts of Calais very rarely record information pertaining to repairs carried out to guns. TNA, E 159/247, recorda, Trinity, rot. 7.

61 TNA, E 101/294/21.

62 Lorraine C. Attreed, ed., *The York House Books: 1461–1490*, volume 2 (Gloucester: Alan Sutton for Richard III & Yorkist History Trust, 1991), pp. 307, 617.

63 LMA, COL/CC/01/01/009, f. 43v.

64 SRO, SC 5/1/15, f. 26v.

iron and constructing a new stock with timber, before the gun was finally 'dressed'.[65] The adoption of individual carts or carriages for guns, which took place in the mid-fifteenth century, meant that significant resources had to be allocated for their upkeep, as can be seen from the accounts for the 1497 expedition to Scotland.[66] Objects made out of organic materials, such as the leather used to make horse harnesses, were particularly susceptible to damage, especially when left in storage for long periods of time, as can be seen from a reference in 1497 to harnesses that were 'hurt and perished by rats and other vermin'.[67]

There is limited evidence that some pre-emptive measures were carried out in an attempt to prevent damage to ordnance. This included hiring painters to treat equipment with paint, which can act as a preservative. Two men were hired to paint six fowlers and fifty-four serpentines along with their carts at the Tower of London in 1471–2.[68] Ten years later, a John Bulle of Calais was hired to paint serpentines and fowlers located in the walls, bulwarks and murderers of the town.[69] Grease, such as tallow, was also used as a means of preventing damage to carts used to transport ordnance.[70] Safety checks were carried out on newly constructed guns, which were often test fired or 'proved' to ensure that they were structurally sound prior to their use on campaigns. Mile End, located just outside the eastern boundary of the city of London, was a favoured location for such tests.[71] Richard Cely the Younger provides a description of a newly cast gun that shattered when tested at Mile End on 27 March 1482: referring to 'the grehyt new gone of brasse shott at Mylezeynde at whos mad in the Towyr and hyt braste awll to pessys'.[72] The presence of certain named guns in inventories over the course of many years, however, suggests that it was possible to provide long-term maintenance for these weapons. For example, a large iron bombard called the *Saint Paul* and another of bronze called the *Dame Anable* were first listed in the accounts of the treasurer of Calais in 1434–6 and were still present more than fifty years later in 1488.[73]

65 BL, Egerton MS 2090, f. 152r.

66 TNA, E 36/8, ff. 26r, 27r, 29r, 30v, 32v, 39r, 72v–73r, 75r.

67 Ibid., f. 95r.

68 TNA, E 101/294/19.

69 TNA, E 101/55/4, f. 22r.

70 TNA, E 36/8, f. 77v.

71 For examples of this see TNA, E 36/8, ff. 51v, 54r.

72 Henry Elliot Malden, ed., *The Cely Papers: Selections from the Correspondence and Memoranda of the Cely Family, Merchants of the Staple, A.D. 1475–1488* (London: Longmans, Green, and Co., 1900), p. 88.

73 TNA, E 364/72, rot. E; E 364/119/36, rot. D.

AMMUNITION

Different types of projectiles were used for guns throughout the period made out of lead, stone and iron.[74] The earliest English firearms of the mid-fourteenth century were small in size and were designed for firing lead shot; this was true of the ten guns sent from the privy wardrobe for the siege of Calais in 1346–7.[75] Lead was heated in pans until it turned into a liquid state, and then it was poured into moulds of bronze or stone to cast bullets. There is evidence that rings of iron could be used to compress the pellets if they were too large in size.[76] The finished bullets were then stored in chests of wood or baskets of fibre, and canvas was used to cover the latter.[77] This work was carried out by men described as 'operators' in the privy wardrobe accounts, but by the late fifteenth century was undertaken by gunners.[78] The growing importance of small calibre guns, such as serpentines and handguns, from the mid-fifteenth century onwards, led to an increase in the quantities of lead shot cast. Different types of mould were used to make bullets for serpentines, falcons and hackbuts for the 1497 expedition to Scotland.[79] The majority of projectiles used in the fifteenth century, however, were made out of stones known as 'gunstones'. These were obtained from quarries throughout England, although a preference was shown for obtaining stone from Maidstone for the provision of the royal artillery; for instance 1,214 gunstones were purchased from five masons of the town in 1428.[80] The stones were carved into smaller round balls by masons using hammers, chisels and stone axes, as can be seen in the case of 3,364 gunstones made out of 1,552 pieces of Kentish ragstone by William Warlowe, a mason of Calais in 1434–6.[81] Gunstones of different sizes were required for particular types of guns. The artillery constructed in 1453–6 required stones ranging in size from four to twenty-two inches in diameter, and moulds of oak were used by the masons to ensure consistency.[82]

74 Note that there are also a few references to quarrel guns in the late fourteenth–early fifteenth century but very little information about them survives. For an example see TNA, E 403/567, m. 1.

75 Tout, 'Firearms in England in the Fourteenth Century', pp. 688–9.

76 TNA, E 364/119/36, rot. B; Tout, 'Firearms in England in the Fourteenth Century', pp. 696–700; E 101/198/13, ff. 48r, 92r; E 36/8, ff. 72r, 91v.

77 TNA, E 101/55/4, f. 21v; E 36/8, f. 67v.

78 Tout, 'Firearms in England in the Fourteenth Century', pp. 693–4; TNA, E 36/8, f. 67v.

79 TNA, E 36/8, ff. 58v, 66r.

80 Spencer, 'The Provision of Artillery for the 1428 Expedition to France', p. 185; *CPR 1416–1422*, pp. 134, 262; TNA, E 403/636, m. 18; E 101/294/19.

81 TNA, E 101/61/9 f. 7r; E 36/8, f. 68v; E 364/72, rot. E.

82 Spencer, 'The Lancastrian Armament Programme of the 1450s and the Development of Field Guns', p. 70.

Iron was also used as a projectile for guns. The earliest record for its usage occurs in 1438, when a team of German gunners were paid to test fire different ammunition including one gunstone made out of iron.[83] There is very little evidence for this type of shot prior to the reign of Edward IV, although the citizens of Coventry purchased thirteen bullets of iron and lead for two large guns in 1451.[84] From the 1470s onwards, this type of ammunition was purchased for the Calais garrison from the Low Countries. In 1471–2, 106 lead shot were acquired in addition to 260 for large serpentines and 172 'hail shot' in 1485.[85] By 1490, gunstones of iron were being manufactured in England as we can see from payments made by Richard Guildford, master of the ordnance, to workers in Suffolk and Sussex.[86] This was almost certainly as a result of the introduction of the blast furnace to England at this time, which meant that iron could be decarburised to produce cast iron.[87] It rapidly became the preferred type of projectile, and most of the ammunition provided for Henry VII's expedition to France in 1492 was iron shot.[88] Four years later, cast iron shot was cast at Newbridge in Ashdown Forest in Sussex as part of the preparations for the Scottish campaign of 1497.[89]

GUNPOWDER

The adequate provision of gunpowder was a crucial aspect of ensuring that guns could be used when required. Gunpowder, also referred to as black powder, is created from mixing three components together: saltpetre (potassium nitrate), sulphur and charcoal.[90] Of these, charcoal could be easily procured from England, with willow and lime trees often being used, but the other two substances had to be imported from abroad.[91] Sulphur could be obtained from southern Italy, but saltpetre, the most important ingredient, was mainly supplied from India via the

83 TNA, E 101/503/7.

84 Dormer, *The Coventry Leet Book*, p. 262.

85 The latter was a type of shot designed for targeting anti-personnel see Glenn Foard and Anne Curry, *Bosworth 1485: A Battlefield Rediscovered* (Oxford: Oxbow Books, 2013), p. 149; TNA, E 101/197/14; E 364/119/36, rot. D.

86 TNA, E 36/124, f. 85v; E 405/78, m. 34v.

87 Brian Awty and Christopher Whittick., 'The Lordship of Canterbury, Iron-Founding at Buxted, and the Continental Antecedents of Cannon-Founding in the Weald', *Sussex Archaeological Collections*, 140 (2002), 71–81 (p. 71).

88 TNA, E 36/15, f. 8r.

89 TNA, E 36/8, ff. 38r, 54r.

90 David Cressy, *Saltpeter: The Mother of Gunpowder* (Oxford: Oxford University Press, 2013), p. 10.

91 TNA, E 403/735, m. 5; E 364/95, rot. E; Tout, 'Firearms in England in the Fourteenth Century', pp. 696–700.

Mediterranean, although plantations were also established in northern Europe in the late fourteenth century.[92] The increase in the supply of saltpetre, due to imports from the Continent, appears to have led to a significant fall in the price of gunpowder in England in the early fifteenth century.[93] Merchants of London, such as mercers and grocers, played an important role in supplying saltpetre to the crown. The most notable example, Thomas Chalton, received at least £661 9s. 2d. between the years 1417 and 1422.[94] Saltpetre was also purchased from Iberian or Italian merchants. Thus 8,100lbs were procured from Catalonia in 1421 and twenty-four barrels from Palermo in 1491.[95] Supplies for the Calais garrison were partly obtained from the Low Countries, as were supplies for some expeditions to France, as can be seen from the despatch of three gunners to Flanders to procure provisions in 1430.[96] Efforts were made to establish a domestic industry in the reign of Henry VII. A commission was given to a certain James Hede in 1492 to take houses, lands, wood, coals and workers to make saltpetre, but the crown remained heavily dependent upon imports into the sixteenth century.[97]

Gunpowder was generally purchased from merchants and gunners. John Nicholl, a grocer of London, was a prominent supplier during the reign of Edward IV.[98] An insight into how gunpowder was created can be seen from payments made at Southampton in 1456–7. The town's gunner, John Bronne, assisted by three other men, was paid for four days work making gunpowder, in addition to a labourer who was employed for two days to beat the charcoal. They were supplied with 91lbs of saltpetre by an Angell Aldebrand, most likely an Italian merchant, 56lbs of sulphur by the mayor, Walter Clerk, as well as 84lbs of charcoal, four sieves and six white leather bags.[99] Other tools used for grinding, mixing and weighing gunpowder included mortars, pestles, stampers, troughs, tubs, pans, mills and balancing scales.[100] The finished product was then placed in barrels, firkins, hogsheads or pouches, which were kept in storehouses, towers or other buildings. Special 'close' carts were used for keeping the powder safe whilst it was being moved during campaigns,

92 Cressy, *Saltpeter*, p. 12; Bert S. Hall, *Weapons and Warfare in Renaissance Europe* (London: The John Hopkins Press Ltd, 1997), pp. 74–5.

93 See Appendix D.

94 TNA, E 403/567, m. 1; E 403/595, m. 21; E 403/630, m. 2; E 403/636, m. 14; E 403/639, m. 2; E 403/642, m. 6; E 403/825, m. 5.

95 TNA, E 403/649, m. 5; C 1/108, no. 61; E 403/644, m. 3; E 405/566, f. 63r.

96 TNA, E 403/619, m. 2; E 364/69, rot. Q dorse; E 101/197/14; E 101/197/20, f. 3r.

97 *CPR 1485–1494*, p. 395; Cressy, *Saltpeter*, pp. 42–7.

98 TNA, E 403/777, m. 5; E 101/195/19, f. 5r; E 101/195/14, f. 2r; E 403/823, m. 3; E 403/825, m. 9; E 403/841, m. 8; E 405/53, m. 3r; E 405/57, m. 1r; E 405/77, mm. 4d, 6d; E 405/78, mm. 1d, 6d; E 36/124, ff. 45v, 77r, 78v; DL 37/33.

99 SRO, SC 5/1/8,, ff. 12r–12v, 14v.

100 TNA, E 101/198/13, ff. 80r, 81r, 89r; E 364/119/36, rot. B; Spencer, 'The Lancastrian Armament Programme of the 1450s and the Development of Field Guns', pp. 63–4.

doubtless to minimise the risk of accidental explosions.[101] Particular types of gunpowder designed for the use of specific types of guns emerged throughout the course of the fifteenth century.[102] These were created using different proportions of ingredients, as can be seen from the detailed purchases made for the Calais garrison in 1485. On this occasion, saltpetre comprised 70 per cent of the touch powder, 66 per cent of the serpentine powder, 65 per cent of the hackbut powder and 50 per cent of the bombard powder.[103] By the 1480s, gunpowder was also kept in a standard format, with a barrel containing 200lbs and a last containing 2,400lbs.[104]

CONCLUSION

The adoption of gunpowder weapons in medieval England was assisted by the ease with which traditional techniques employed by smiths, brasiers or bell-founders could be adapted for gun-making. This also meant that firearms could be repaired, transported and supplied with ammunition by utilising existing methods. The evidence that remains about named guns also shows that it was possible to keep firearms in active service for long periods of time, with efforts being made to maintain and test fire the weapons to ensure that they were safe to use. Yet major changes took place throughout the course of the fifteenth century, prompted by developments in the production of gunpowder, transportation and metallurgy. The first two decades of the century saw a shift from predominantly using bronze to using cheaper wrought iron for gun-making, as well as a dramatic fall in the price of gunpowder. These changes meant that it became feasible to employ firearms on an extensive scale in siege warfare without incurring ruinous expenses in the process. Another important innovation was the use of individual carriages for guns in the late 1440s, which soon led to the use of field guns on the battlefield in the 1450s. The most significant developments, however, took place in the 1480s and 1490s, with the introduction of new types of bronze firearms, gunpowder recipes, cast iron projectiles and the foundation of a domestic saltpetre industry. English artillery therefore changed markedly over the course of the century, as the increased cost of ordnance, notably due to the resources needed to maintain sizeable numbers of horses and carts, transformed the organisation of war.

101 TNA, E 101/55/4, f. 14r; E 101/198/13, ff. 81r–82v; E 36/8, ff. 5r, 17v, 19v, 103r; CCA, CC/F/A7, f. 10r; *CPR 1461–1467*, p. 23.

102 Serpentine powder is first recorded in the accounts for 1461–2, hackbut and bombard powder in 1475, and touch powder in 1481. TNA, E 364/72, rot. E; E 364/96, rots. B, C; E 101/198/13, f. 6r; E 364/119/36, rot. C.

103 TNA, E 101/55/11.

104 TNA, E 101/198/13, f. 14v; E 36/8, f. 2v.

Conclusion

Multiple factors were responsible for the adoption and development of firearms by the English crown and urban communities. The initial spread of guns was aided by the ease of their manufacture, with traditional techniques employed by metal workers utilised for the construction of these new weapons. In addition, the evidence provided by both royal and town records show that gunpowder weapons were often acquired as a response to external threats. This can be clearly seen in the case of the Calais garrison, where increases in the number of guns possessed by the garrison often took place during periods of tensions, such as in 1385–6, 1436 and 1450–1.[1] Similarly, many towns acquired guns when they were threatened with attack, particularly in the 1450s and 1480s. As Chapter 7 on Southampton demonstrates, though, some settlements were more willing to invest in the new technology than others. Developments in artillery fortifications generally followed a similar pattern, with the decision to build bulwarks only being undertaken by the aldermen of some English towns in the second half of the fifteenth century, despite the construction of a bulwark near Southampton during the reign of Henry V. Likewise, investment in the defences of the Pale of Calais tended to be significantly increased when the territory was threatened with attack. This meant that there was often a delay between technological developments and the adoption of new gun types or fortifications.

By contrast, royal influence played a more active role in encouraging the acquisition and development of firearms. This stemmed in part from the direct interest of most English kings in gunpowder weapons, with even the unwarlike Henry VI being witness to the testing fire of guns. The larger types of firearms, particularly the bombards, were prestigious and expensive weapons, often named after the king or other important individuals, and they were sometimes mentioned by name in chronicles. This meant that the deployment of royal guns became a symbol and expression of a sovereign's wealth, power and prestige, whether to impress rival rulers, as with Edward IV's expedition to France in 1475, or domestic rebels, such as with the Cornish rising of 1497. Yet the reputation of these weapons ultimately derived from their effectiveness in warfare, and

1 See Chapter 4, Figure 2.

English kings were keen to acquire the best firearms to improve their prospects in offensive and defensive military operations. The re-establishment of a fleet of royal ships owned by the crown, some of which were heavily armed with guns, was a direct result of Edward IV's desire to possess a powerful naval force. Similarly, the formation of a mobile arsenal in Calais meant that he had large quantities of modern ordnance to equip his expeditionary armies. Royal support was also offered to a small number of coastal towns in the second half of the fifteenth century, due to the high cost of gunpowder weapons and their perceived vulnerability to attack. This played an important part in strengthening the defences of these communities, and English kings were eager to encourage urban authorities to invest in firearms. The recruitment of foreign gunners and specialists from the Low Countries and elsewhere also meant that royal involvement was crucial in the transmission of new ideas and techniques from the Continent. This was especially important as most, if not all, technological advances in gun-making came from abroad.

Taken collectively these developments would appear to warrant the description of a military revolution, as warfare was radically different in 1500 to what it had been in 1400, due to significant changes in gunpowder weapons. Yet it is difficult to come to a firm conclusion on this question as there is no clear definition of what constitutes a military revolution amongst historians. Three major periods of technological change can be clearly identified during the course of the century: the emergence of the bombard in the early 1400s, the adoption of the serpentine in the 1450s, and the invention of new bronze guns firing iron shot in the reign of Henry VII. Each of these developments ultimately had a major impact on the way that the English used firearms and could well be described as revolutionising warfare. Yet change was often accumulative, as can be seen in the case of the serpentine, which only reached its full potential in the 1470s, twenty years after it was first adopted, whereas improvements to existing models occurred gradually throughout the period. This was not a consistent process, however, as the rate of change notably increased in the second half of the century, due to the heavy investment of the governments of Edward IV and Henry VII in firearms.

The types and ways that guns were used by the English changed significantly throughout the course of the fifteenth century. At the beginning of the period, firearms were either very large or small and were primarily used by the crown for attacking or defending fortifications. By the close of the century, however, guns had changed almost beyond recognition, with the emergence of many different types of weapons, which were widely present in ships, town arsenals and on the battlefield. These technological developments meant that firearms came to occupy an integral role in all aspects of warfare by the third quarter of the fifteenth century. A consequence of this development was that greater resources needed to be allocated to the administration, procurement, transportation and operation of gunpowder artillery. The English state, unlike some of its European rivals, did not feel the need to establish a permanent standing army in the late fifteenth century. Yet it was considered necessary to set up a permanent ordnance office, served by a small number of officials and gunners, for the effective deployment of firearms in military operations. These changes meant that the costs of both waging war and maintaining defences increased markedly, which placed a heavy burden, in particular, on urban communities, with royal subsidies provided for key coastal settlements. The English experience of warfare was therefore transformed in the fifteenth century due to developments in the technology of firearms.

Yet the speed and adoption of technological change did not take place at a consistent rate; most progress occurred in the second half of the period. The firearms of the fourteenth century were small and primarily used for the defence of royal castles or towns, in part due to the high price of gunpowder.[2] This changed during the reign of Henry IV with the emergence of very large guns, which were given their own individual names. These weapons were capable of causing significant damage to the defences of enemy settlements and thereby were able to influence the outcome of sieges. Henry V made extensive use of firearms during his conquest of Normandy, which was made possible, at least in part, by a marked fall in the price of gunpowder. During this period a clear distinction emerged between large bombards, large fowlers and very small guns, such as fowlers and handguns. The design of these weapons meant that they were most suitable for use in siege warfare. The bombards were capable of firing heavy shot, whereas the fowlers could maintain a high rate of fire, albeit at comparatively short range. A sizeable number of royal ships were armed with small numbers of guns during the reigns of the Lancastrian kings, but these appear to have had very little, if any, impact on warfare at sea. At this time a shift also occurred in gun production methods. Most guns were now made from wrought iron as opposed to bronze. This occurred primarily due to the lower cost of wrought iron guns, which also had the advantage that they could be easily repaired, whereas the latter had to be recast if damaged. The growing effectiveness of firearms is likely to have been a factor in the decision to improve the defences of the Pale of Calais, in the 1420s and 1430s, with the addition of bulwarks. Nevertheless, these weapons appear to have had a relatively limited impact on the conduct of warfare, as they were, on the whole, improvements on existing models. Furthermore, the aldermen of English towns rarely felt obliged to invest in firearms at this time, with Southampton's commitment to the new technology being exceptional.

The most significant development in the fifteenth century was the adoption of the serpentine in the 1450s. This new type of gun was ideally suited to targeting enemy personnel at range with some accuracy due to its long shape and narrow bore. The use of the serpentine ultimately transformed the conduct of warfare; for the first time, guns could be used effectively in sieges, in naval warfare and on the battlefield. One of the earliest applications of this new weapon was in the field;

2 See Appendix D.

serpentines were regularly deployed by English armies from the 1450s onwards. This led to a significant change in the methods used to transport ordnance. There was a shift away from large carts containing multiple pieces being moved by oxen to guns being mounted on their own individual carriages pulled by horses. Serpentines were also acquired for the defence of royal fortifications, principally in the Pale of Calais, as well as in urban arsenals. Existing types of firearms continued to increase in size at this time; the heaviest known English bombard of the century, the *Goodgrace*, weighed as much as 22,400lbs in 1456.[3] The decade also saw the adoption of guns by a sizeable number of towns in England, principally coastal settlements which were vulnerable to raids. This prompted many of these urban communities to adapt their fortifications to the threat and potential of gunpowder artillery, often through the construction of bulwarks.

Firearms continued to develop and diversify during the reign of Edward IV, particularly after 1471, with the emergence of new gun types, such as pot guns, bombardelles, curtows and bastard guns. The surviving views of the weaponry in the fortifications of Calais in the 1470s and 1480s demonstrate that guns had evolved to serve a variety of tactical roles. The long-ranged serpentines were often deployed on walls, whereas shorter-ranged guns, such as fowlers, were primarily placed in murderers or above gatehouses. Handguns also began to play a more prominent role in major expeditions, urban arsenals and in the defence of Calais.[4] They were only used in relatively small numbers in the fifteenth century, however; for instance only 111 hackbuts were transported to Scotland from Calais in 1481. The handgun was therefore a niche weapon used in specific situations during the fifteenth century, such as in sieges or on-board vessels, which meant that it did not supplant, or even threaten, the traditional role of the longbow in English warfare. By contrast, naval operations were radically changed by the invention of the miche, most likely in the 1470s, which meant that serpentines could be used in large numbers on ships, primarily in their castles, thus dramatically improving the firepower of those vessels. Serpentines were intended primarily for targeting enemy personnel, in conjunction with traditional weapons such as the longbow, yet the development of the miche was a necessary prerequisite to enable their use on the heavily armed ships of the sixteenth century. The second reign of Edward IV also saw a significant expansion in the number of guns possessed by the crown, as can be seen by the large quantities of weaponry kept at the Pale of Calais.[5] This prompted the adoption of new techniques to maximise the efficiency of firearms; new varieties of gunpowder were developed for the specific use of different guns. Advances in artillery fortifications kept pace with these changes, as can be seen from the extensive remodelling of the defences of Guînes Castle from the 1460s

3 See Appendix A.

4 See Figures 3 and 4.

5 See Figure 2.

until the 1480s. The Calais evidence also indicates that the role of the gunner was changing in this period, with a growing emphasis on the operation as opposed to the production of guns.

The fastest rate of technological change occurred during the reign of Henry VII. This period saw the emergence of ordnance which differed notably from that used during the reign of Edward IV. This can be seen by comparing the guns sent for the Scottish expedition of 1481 with those used in the Boulogne campaign of 1492. In 1481, the guns were mostly wrought iron bombards, pot guns, hackbuts and serpentines, which fired gunstones and lead shot. By contrast, the artillery sent to France in 1492, just eleven years later, consisted of bombardelles, curtows, demi-curtows, serpentines, falcons and hackbuts, almost all constructed of bronze, firing a mixture of iron shot, gunstones and lead shot. At least part of the variance in gun types may have reflected efforts to reform the terminology used to describe different pieces, further evidence for which can be seen with Robert Clifford's accounts for the 1497 expedition to Scotland, with specified weights for the size of shot and gunpowder used to fire different categories of guns. The improvements in the quality of gunpowder artillery came at a cost, however, as can be seen by the extensive quantities of equipment allocated to the land and naval campaigns of the 1490s. This necessitated changes to the administration of the ordnance. In particular, the status of the master of the ordnance increased. Therefore the artillery available to Henry VII at the end of the century was radically different to what had been available to his Lancastrian predecessor Henry IV.

These conclusions broadly correspond with the trends identified by previous scholars for the progression of gunpowder artillery in Europe, notably the development of longer guns and the shift to primarily constructing guns out of bronze, as opposed to wrought iron.[6] They also indicate that for much of the fifteenth century, English kings were able to keep up with the new types of firearms used on the Continent, notably those possessed by England's main rivals, the French and Burgundian states. This means that traditional interpretations of English technological backwardness can be firmly rejected.[7] Yet it would be wrong to assume that the English experience was exactly the same as that of Continental rivals. Despite early evidence for the use of guns by the English, such as on ships in the 1330s or at the Siege of Calais in 1346–7, they were at first relatively slow to adopt gunpowder weapons in a meaningful way. Guns were not consistently issued to supply royal castles until the 1360s and towns only began to acquire firearms in the last quarter of the century. By contrast, guns are listed as being present in the

6 Bert S. Hall, *Weapons and Warfare in Renaissance Europe* (London: The John Hopkins Press Ltd, 1997), pp. 87–92, 210.

7 For example see Jonathan Sumption, *The Hundred Years War: Volume IV, Cursed Kings* (London: Faber, 2015), pp. 418–19.

arsenals of towns and castles and mentioned as being employed in siege warfare in France and Italy from the 1330s onwards.[8] Similarly, large guns, later known as bombards, were first used in northern Europe in the 1370s, notably at the siege of Saint-Sauveur-le-Vicomte in 1375, but the English did not adopt this type of gun until the first decade of the fifteenth century.[9] This delay can be attributed to England's reliance on *chevauchées* during the fourteenth-century phase of the Hundred Years' War.

The situation was changed in the early fifteenth century due to the efforts made by Henry IV and Henry V to acquire the latest gunpowder weapons for use in offensive siege warfare. This allowed them to build up an artillery train that could rival that of their Continental adversaries. Henry V's appointment of a master of the ordnance to administer the royal guns in 1413 was contemporary with the decision by John the Fearless, duke of Burgundy, to reform his artillery two years later.[10] Similarly, the English adoption of fowlers coincided with their first mention in French records.[11] New artillery fortifications known as bulwarks, first recorded in France in the first decade of the fifteenth century, were also constructed by the English following Henry V's expeditions to the Continent.[12] However, the breakdown of the Burgundian alliance led to a period of stagnation. New types of guns with long barrels, such as *crapaudeaux* and serpentines, came into use on the Continent in the 1430s and 1440s.[13] Yet the English did not employ the latter until the 1450s, when the shock of defeat in France prompted the crown and towns to

8 Philippe Contamine, *War in the Middle Ages*, trans. Michael Jones (Oxford: Blackwell, 1984), p. 139; Emmanuel de Crouy-Chanel, *Canons Médiévaux: Puissance du Feu* (Paris: Rempart, 2010), pp. 35–40. It is worth noting that other European states, such as the city state of Dubrovnik and the kingdom of Scotland, only began to use gunpowder weapons for the first time in the 1370s and 1380s. Djurdjica Petrovic, 'Fire-arms in the Balkans on the Eve of and After the Ottoman Conquests of the Fourteenth and Fifteenth Centuries', in *War, Technology and Society in the Middle East*, ed. V. J. Parry and M. E. Yapp (London: Oxford University Press, 1975), pp. 164–94 (168–72); David H. Caldwell, 'The Scots and Guns', in *England and Scotland in the Fourteenth Century: New Perspectives*, ed. Andy King and Michael A. Penman (Woodbridge: Boydell Press, 2007), pp. 60–72 (p. 72).

9 Contamine, *War in the Middle Ages*, pp. 140–1; Crouy-Chanel, *Canons Médiévaux*, p. 44.

10 Richard Vaughan, *John the Fearless: The Growth of Burgundian Power* (London: Longmans, 1966), pp. 150–1; Robert D. Smith and Kelly DeVries, *The Artillery of the Dukes of Burgundy 1363–1477* (Woodbridge: Boydell Press, 2005), p. 19.

11 Note that they are called *veuglaires* in French records. Crouy-Chanel, *Canons Médiévaux*, p. 54.

12 Contamine, *War in the Middle Ages*, p. 204.

13 Ibid., p. 142; Crouy-Chanel, *Canons Médiévaux*, p. 77; Smith and DeVries, *The Artillery of the Dukes of Burgundy 1363–1477*, pp. 221–4.

update and modernise their stocks of artillery. Thereafter the development of the technology in England broadly kept pace with those of her main north European rivals. This can be seen with new types of guns, such as hackbuts, which were purchased from continental gun-makers for the Calais garrison almost as soon as they appeared on the market.[14] Similarly, the English were quick to adopt the use of new French bronze guns in the 1490s.[15] The construction of new specialised artillery fortifications at Calais were also contemporary with the works carried out by Louis XI at Perpignan, Dijon and elsewhere.[16]

Yet an important difference with other European powers was the relationship between the state and the towns. Rulers such as the dukes of Burgundy were dependent upon urban arsenals for supplying guns for their military campaigns, with the wealthy settlements of the Burgundian Low Countries regularly asked to send ordnance to furnish the ducal artillery train.[17] By contrast, the opposite situation prevailed in England. English kings were keen to encourage towns to invest in guns and artillery fortifications, with ordnance even sent to furnish the defences of key places. This difference was due, at least in part, to varying military, political and economic circumstances. The wealthy settlements of the Low Countries, as well the imperial free cities, often had a degree of self-rule, but were conscious of the threat posed to them by neighbouring powers. This meant that they needed to invest heavily in fortifications and weaponry. Cologne had 348 guns in 1468, Strasbourg 585 in 1476, Nuremberg 2,230 in 1462, and Ghent had 189 in 1456 and 486 in 1479.[18] Most English towns, except for London, were much smaller in population and economic size and faced a considerably lower risk of being besieged. Therefore, there was less incentive to invest heavily in artillery and English kings were forced to rely on their own resources for acquiring guns. Furthermore, the development and use of gunpowder weapons varied considerably throughout the Continent. The Hussite use of *Wagenburg* tactics became a mainstay of warfare in central Europe in the fifteenth century, yet these methods were not employed in western Europe.[19] Other states were relatively slow in adopting gunpowder weapons. The Iberian kingdoms of Castile and Aragon made only limited use of guns in warfare prior to the Granada War of 1482–92 and gunpowder weapons only became an essential part of Scottish military operations during the reign of

14 Hackbuts are first recorded as being present in the arsenal of the Calais garrison in 1469. TNA, E 364/119/36, rot. C. Hackbuts are recorded as being present in French and German towns in the 1470s, Crouy-Chanel, *Canons Médiévaux*, p. 106.

15 TNA, E 36/124, f. 39.

16 Crouy-Chanel, *Canons Médiévaux*, p. 110.

17 See Smith and DeVries, *The Artillery of the Dukes of Burgundy 1363–1477*.

18 Contamine, *War in the Middle Ages*, p. 148.

19 Hall, *Weapons and Warfare in Renaissance Europe*, pp. 107–9.

James I (1437–60).[20] Therefore, the English experience was unique as technological change and innovation across the Continent was influenced by differences in military, political and economic circumstances.

20 Weston F. Cook, 'The Cannon Conquest of Nasrid Spain and the End of the Reconquista', *The Journal of Military History*, 57 (1993), 43–70 (pp. 48–50); D. H. Caldwell, 'Guns in Scotland: The Manufacture and Use of Guns and their Influence on Warfare from the Fourteenth Century to c.1625' (unpublished doctoral thesis, University of Edinburgh, 1982), pp. 126–7.

Appendix A

Weights of Guns[1]

Date	Quantity	Type of gun	Material	Name of gun	Weight (lbs)	Average weight (lbs)
1382–8	4	Large gun	Bronze		700	175
1382–8	7	Gun	Unknown		484	69
1382–8	5	Gun	Unknown		1422	284
1382–8	47	Large gun	Unknown		15,967	340
1382–8	1	Large gun	Unknown		665	665
1382–8	9	Pellet gun	Bronze		383	42.5
1382–8	1	Large gun	Bronze		210	210
1382–8	1	Large gun	Unknown		242	242
1382–8	1	Gun	Unknown		163	163
1400–15	1	Gun	Iron	*Fowler*	1,100	1,100
1400–15	1	Gun	Bronze		200	200
1400–15	1	Large gun	Bronze	*The George*	3675	3675
1400–15	1	Gun	Bronze	*Messenger*	4480	4480
1400–15	2	Gun	Bronze		500	250
1400–15	1	Gun	Bronze	*Gobette*	1,100	1,100
1400–15	1	Large gun	Iron	*King's Daughter*	5,600	5,600
1407–10	1	Gun	Iron		2,027	2,027
1428	2	Fowler	Bronze		168	84

[1] T. F. Tout, 'Firearms in England in the Fourteenth Century', *The English Historical Review*, 26 (1911), 666–702 (pp. 691, 696–700); TNA, E 364/49, rot. C; E 101/51/27; E 364/66, rot. C; E 364/69, rot. B dorse; E 364/90, rot. D; E 404/71/1, no. 60; E 364/103, rots. E, G; E 159/249, recorda, Hilary, rots. 12d–13; E 364/106, rot. I; E 101/55/4.

Date	Quantity	Type of gun	Material	Name of gun	Weight (lbs)	Average weight (lbs)
1428	3	Large gun	Iron		15,500	5,167
1428	1	Large gun	Iron		5,350	5,350
1428	3	Large gun	Iron		14,800	4,933
1430	1	Large gun	Iron	*Henry*	6,780	6,780
1430	1	Large gun	Iron	*Crown*	7,122	7,122
1431–2	5	Pellet gun	Iron		220	44
1431–2	3	Gun	Iron		216	72
1431–2	1	Gun	Iron		1,220	1,220
1432–4	1	Gun	Bronze		3,904	3,904
1453–6	1	Bombard	Iron	*Goodgrace*	22,400	22,400
1453–6	1	Bombard	Iron	*Henry*	19,040	19,040
1453–6	1	Bombard	Iron	*Crown*	7,840	7,840
1453–6	4	Fowler	Iron		7,840	1,960
1453–6	4	Serpentine	Iron		840	210
1453–6	25	Cart gun	Iron		8,105	324
1457	26	Serpentine	Iron		5,200	200
1467–8	1	Serpentine	Bronze		700	700
1467–8	1	Large gun	Iron		1988	1988
1467–8	1	Serpentine	Iron		858	858
1468–9	1	Serpentine	Iron		907	907
1468–9	1	Serpentine	Iron		230	230
1468–9	1	Fowler	Iron		112	112
1468–9	2	Fowler	Iron		717	358
1468–9	1	Serpentine	Iron		378	378
1468–9	1	Serpentine	Iron		529	529
1472	1	Fowler	Iron		233	233
1472	1	Serpentine	Iron		208	208
1472	1	Fowler	Iron		340	340
1472	1	Serpentine	Iron		585	585
1472	1	Fowler	Iron		336	336
1472	1	Fowler	Iron		560	560
1472	1	Fowler	Iron		450	450
1472	1	Fowler	Iron		440	440

Date	Quantity	Type of gun	Material	Name of gun	Weight (lbs)	Average weight (lbs)
1472	9	Cart gun	Iron		2,400	267
1472	2	Serpentine	Iron		485	242
1472	3	Serpentine	Iron		340	113
1472	3	Serpentine	Iron		200	67
1472	1	Serpentine	Iron		280	280
1472	1	Serpentine	Iron		180	180
1472	1	Pot gun	Iron		2,000	2,000
1472	1	Serpentine	Iron		4,200	4,200
1472	1	Large fowler	Iron		3,100	3,100
1472	1	Fowler	Iron		1,600	1,600
1472	1	Long fowler	Iron		1,900	1,900
1472	1	Serpentine	Iron		2,100	2,100
1472	1	Serpentine	Iron		1,400	1,400
1472	1	Serpentine	Iron		750	750
1472	1	Serpentine	Iron		750	750
1472	1	Fowler	Iron		1,100	1,100
1472	1	Fowler	Iron		600	600
1472–3	14	Large fowler	Iron		5,707	408
1472–3	5	Serpentine	Iron		2,078	416
1472–3	1	Gun	Iron		1,324	1324
1476–7	15	Serpentine	Bronze		4,225	282
1477–8	20	Serpentine	Iron		8,204	410
1477–8	9	Fowler	Iron		2,100	233
1478–9	4	Serpentine	Bronze		2,260	565
1479–80	3	Fowler	Iron		697	232
1480–1	4	Serpentine	Iron		2,946	736.5
1481–2	9	Serpentine	Iron		8,896	988
1481–2	1	Long serpentine	Iron		1,725	1,725
1483–4	15	Serpentine	Iron		8,120	541

Appendix B

Prices of Guns[1]

Date	Quantity	Type	Material	Price	Price per lb
1382–8	4	Large gun	Bronze	£11 13s. 4d.	4
1382–8	7	Gun	Unspecified	£8 1s. 4d.	4
1382–8	5	Gun	Unspecified	£20 14s. 9d.	3.5
1382–8	47	Large gun	Unspecified	£266 2s. 4d.	4
1382–8	1	Large gun	Unspecified	£12 5s. 8d.	4.5
1382–8	9	Pellet gun	Bronze	£6 19s. 8d.	4
1382–8	1	Large gun	Bronze	£3 10s.	4
1382–8	1	Large gun	Unspecified	£4 8d.	4
1382–8	1	Gun	Unspecified	£2 14s. 4d.	4
1428	2	Fowler	Bronze	£4 4s.	6
1428	3	Large gun	Iron	£142 1s. 8d.	2
1428	1	Large gun	Iron	£56 4s.	2.5
1428	3	Large gun	Iron	£148	2.5
1430	1	Large gun	Iron	£56 10s.	2
1430	1	Large gun	Iron	£59 7s.	2
1457	26	Serpentine	Iron	£72 16s.	3
1472	1	Fowler	Iron	£2 15s. 9d.	3
1472	1	Serpentine	Iron	£2 12s.	3
1472	1	Fowler	Iron	£4 5s.	3
1472	1	Serpentine	Iron	£7 6s. 3d.	3
1472	1	Fowler	Iron	£4 4s.	3
1472	1	Fowler	Iron	£7	3

[1] Tout, 'Firearms in England in the Fourteenth Century', pp. 697–8; TNA, E 101/51/27; E 159/249, recorda, Hilary, rots. 12d–13; E 364/106, rot. I; E 101/55/4; E 36/8, f. 76r.

Date	Quantity	Type	Material	Price	Price per lb
1472	1	Fowler	Iron	£5 12s. 6d.	3
1472	1	Fowler	Iron	£5 12s. 6d.	3
1472	9	Cart gun	Iron	£30	3
1472	2	Serpentine	Iron	£4 16s. 3d.	3
1472	3	Serpentine	Iron	£4 5s.	3
1472	3	Serpentine	Iron	£2 10s.	3
1472	1	Serpentine	Iron	£3 10s.	3
1472	1	Serpentine	Iron	£2 5s.	3
1472	1	Pot gun	Iron	£25	3
1472	1	Serpentine	Iron	£52 10s.	3
1472	1	Large fowler	Iron	£38 15s.	3
1472	1	Fowler	Iron	£20	3
1472	1	Long fowler	Iron	£23	3
1472	1	Serpentine	Iron	£26 5s.	3
1472	1	Serpentine	Iron	£17 10s.	3
1472	1	Serpentine	Iron	£9 7s. 6d.	3
1472	1	Serpentine	Iron	£9 7s. 6d.	3
1472	1	Fowler	Iron	£13 15s.	3
1472	1	Fowler	Iron	£7 10s.	3
1476–7	15	Serpentine	Bronze	£74 5s.	4
1477–8	20	Serpentine	Iron	£61 19s. 2 1/2 d.	2
1477–8	9	Fowler	Iron	£14 11s. 8d.	2
1478–9	4	Serpentine	Bronze	£37 13s. 4d.	4
1479–80	3	Fowler	Iron	£5 4s. 6d.	2
1480–1	4	Serpentine	Iron	£22 22s. 1d.	2
1481–2	9	Serpentine	Iron	£71 14s. 3d.	2
1481–2	1	Long serpentine	Iron	£12 16s. 8d.	2
1483–4	15	Serpentine	Iron	£54 2s. 8d.	2
1497	13	Falcons and serpentines	Bronze	£17 10s. 17d.	1.25

Appendix C

Inventory of Ordnance in the Tower of London in 1495[1]

Item	**Quantity**
Bombardelles	2
Great curtows	1
Demi-curtows	8
Serpentines	7
Falcons	31
Small guns	1
Pot guns	1
Hackbuts	15
Serpentines stocked	5
Serpentines unstocked	1
Murderers	1
Other guns	27
Curtows (broken)	1
Demi-curtows (broken)	1
Hackbuts of iron broken	4
Guns called organs	7
Charging ladles for bombardelles, curtows and demi-curtows	18
Charging ladles for serpentines	6
Charging ladles for falcons	25
Gunpowder	7 lasts, 10 barrels
Stampers for curtows and demi-curtows	9
Shot of iron for bombardelles	100

1 TNA, E 36/8, ff. 90r–95r.

Item	Quantity
Shot of iron for curtows, demi-curtows and serpentines	2,070
Shot of iron for small serpentines	153
Shot of stone for bombardelles	70
Shot of stone for curtows, demi-curtows and serpentines	1,580
Shot of lead for serpentines	100
Shot of lead for falcons	1,300
Shot of lead for hackbuts	325
Chains of iron for guns and for the field	70
Boxes of bronze for the great ordnance	5
Dice of iron of sundry sorts	705
Moulds for pellets	4
Charging ladles without staffs for serpentines	6
Wylkyn rammers of iron	2
'Wilkings' of tree with six bars of iron and with twelve 'shevers' of bronze	3
Folding trestles of iron for shooting hackbuts	1
Small charging irons for hackbuts	1
Double plate of iron for charging ladles for the great ordnance	48
Single plates of iron for charging ladles for the small ordnance	49
Rings of iron to compress pellets	5
Square bolts of iron with eleven bolsters for the great ordnance	2
Mortars of iron to beat in gunpowder	2

Appendix D

Prices of Gunpowder[1]

1 Chart showing the average price per lb of gunpowder in pence. Note that this chart is compiled from the surviving evidence, which is far from comprehensive, so needs to be used with caution. Also note that the end date of each accounting period has been used for this chart for the sake of clarity, so for instance 1440–1 has been rendered as 1441. Tout, 'Firearms in England in the Fourteenth Century', pp. 691, 696–700; James Hamilton Wylie, *History of England Under Henry the Fourth, Vol. I. 1399–1404* (London: Longmans, Green and Co., 1884), pp. 187, 233; TNA, E 364/26, rot. A; E 101/404/6; E 101/404/25; E 101/195/11, f. 2a dorse; E 101/195/14, f. 2r; E 101/51/27, m. 2r; E 404/66, nos. 81, 181; E 403/630, m. 4r; E 403/646, m. 1r; E 403/721, m. 9r; E 403/723, m. 14r; E 403/725, m. 15r; E 403/777, mm. 5r, 8r; E 403/791, m. 2r; E 403/810, m. 3r; ESRO, RYE/60/2, f. 60r; RYE/60/3, ff. 15r, 16r, 31r; 82v; HRO, W/E1/35; Mark Stoyle, *Circled with Stone: Exeter's City Walls, 1485–1660* (Exeter: University of Exeter Press, 2003), p. 185; KHLC, LY/2/1/1/1, ff. 53v, 56v, 63b, 77v, 150r, 175v; SA/FAT5, f. 45r; SA/FAT9, f. 5r; BL, MS Add 29616, ff. 132v, 240v, 241r; SRO, SC 5/1/13, f. 33v; SC 5/1/14, f. 7v; SC 5/1/15, ff. 15r; 27r; SC 5/1/18, f. 26r; SC 5/1/24A, ff. 2v, 24a; M. Oppenheim, ed., *Naval Accounts and Inventories of the Reign of Henry VII, 1485–8 and 1495–7, Publications of the Navy Records Society* (London: Naval Records Society, 1896), pp. 13–14.

Appendix E

Total Payments by Towns by Decades (in Pounds)[1]

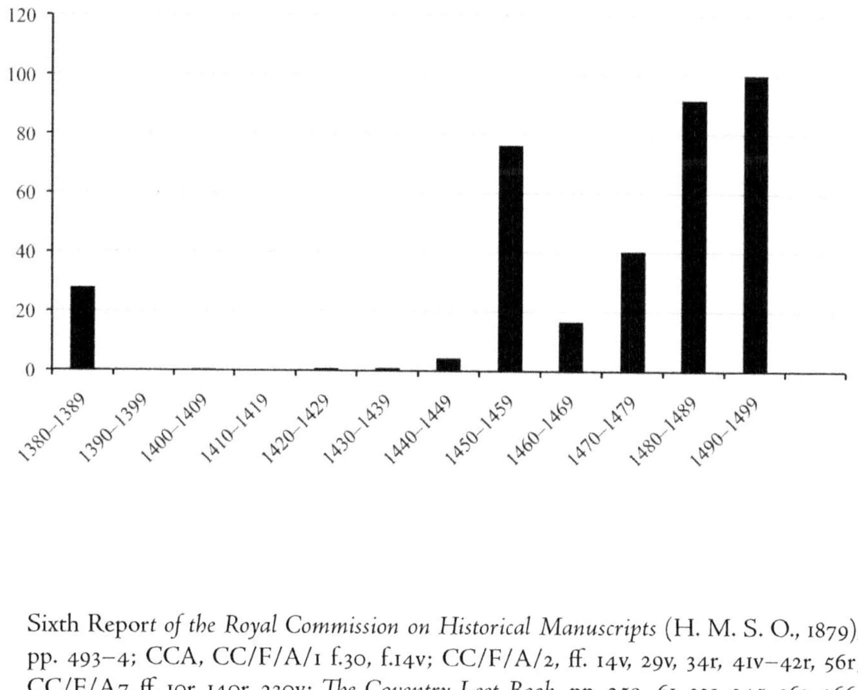

1 Sixth Report *of the Royal Commission on Historical Manuscripts* (H. M. S. O., 1879), pp. 493–4; CCA, CC/F/A/1 f.30, f.14v; CC/F/A/2, ff. 14v, 29v, 34r, 41v–42r, 56r; CC/F/A7, ff. 10r, 140r, 230v; *The Coventry Leet Book*, pp. 259–61, 321, 345, 363, 366; BL, Add MS 29616, ff. 64r–65v, 67r, 83r, 86r, 99v, 132v, 162r, 192r, 208r, 240v–241r, 252r–252v, 262v; Add MS 29617, ff. 10v–11r, 12r, 21r, 22r, 47v–48r, 65v, 74r, 106r, 107r–107v, 129r, 130r, 142r, 201r; Add MS 29810, ff. 19r, 28v; Egerton MS 2090, ff. 10r, 122r, 124v, 136r, 152r–152v; Egerton MS 2107, ff. 4r, 10r, 19v–20r, 29r, 33r, 50r; DHC, *Receivers' Roll Henry VI 31–2 1452–1453; Receivers' Roll Henry VI 39–Edward IV 1 1460–1461*; Stoyle, *Circled with Stone*, pp. 114–16, 131; *Fourth Report of the Royal Commission on Historical Manuscripts* (H. M. S. O., 1874), pp. 434–6; KLBA, KL/C39; KHLC, LY/2/1/1/1, ff. 47r–47v, 50r, 51r, 52r, 53v–54r, 56r–56v, 57v, 61v–62r, 63br, 64r, 67r, 68v, 77v, 104r, 108r, 110r–110v, 111r, 112v, 118r, 119r, 123r, 150r, 152v, 167v–168r, 172r, 175r; NR/FAC2, ff. 51r, 63v; NR/FAC3, ff. 47v–48r, 78r, 102v; William Hudson, ed., *The Records of the City of*

Norwich, volume 2 (Norwich: Corporation of the City of Norwich, 1898), pp. 47–50; PWDRO, W 130, ff. 5r–6r; ESRO, RYE/60/2, ff. 24r, 42r–42v, 43v, 53v–54r, 58v–60r, 66r; RYE/60/3, ff. 16r–16v, 31r, 42r, 46r–46v, 47v, 50r, 79v, 82v, 88, 92v–93r, 113r; RYE/60/4, ff. 200v, 295r–295v, 310v, 317r, 318r, 319r, 322r–322v; KHLC, SA/FAT3; SA/FAT4; SA/FAT5; SA/FAT6; SA/FAT7; SA/FAT8; SA/FAT9; SA/FAT10; BL, Add MS 33511, ff. 7v–8v; SRO, SC 5/1/7; SC 5/1/8; SC 5/1/10; SC 5/1/11; SC 5/1/12; SC 5/1/14; SC 5/1/15; SC 5/1/18; SC 5/1/19; SC 5/1/26; SC 5/1/27b; SC 5/1/28, SC 5/1/29; Gidden, *The Steward's Books of Southampton, from 1428*, volume 1 (Southampton: Cox & Sharland, 1935), pp. 90–3; Harry W. Gidden, ed., *The Book of Remembrance of Southampton, volume 1* (Southampton: Cox & Sharland, 1927), p. 26; HRO, W/E1/35; *York City Chamberlain's Account Rolls 1396–1500* (Durham: Surtees Society, 1980), pp. 96, 130, 137, 139, 150, 153–5, 167, 171, 184, 188; E. W. W. Veale, ed., *The Great Red Book of Bristol*, volume 1 (Bristol: Bristol Record Society, 1931–53), p. 131; DHC, DD 61199a; Richard Peter and Otho Bathurst Peter, *The Histories of Launceston and Dunheved, in the County of Cornwall* (Plymouth: Brendon & Son, 1885), pp. 133–4; Charles Welch, *History of the Tower Bridge* (London: Smith, Elder and Co., 1894), pp. 262–3; LMA, COL/CC/01/01/007 f. 230r.

Appendix F

Extracts from the Accounts of John Hampton 1430–2 – Purchases[1]

Purchased in diverse locations in England and France by command of the king for his aforesaid war within the time of the accounts (1430–2):

- Three large guns of iron, of which one called *Henry*, one called *Hampton* and one called *Crown*
- Sixteen guns of iron called fowlers, with thirty-three moveable chambers, not fixed or stocked
- Four guns of bronze called fowlers, with eight moveable chambers, not fixed or stocked
- 2,128lbs of iron worked
- 6,000lbs of iron not worked
- Twenty chaulders of sea coal
- Four large carts of iron and wood for moving guns, large guns and other things, each with one pair of wheels called trokels
- Four long carts of iron and wood made for use in the country of Picardy
- 1,550 gunstones
- Fifty chains
- Forty staples
- Forty pairs of hooks
- Forty plates of iron for moving oxen
- Eighty collars
- Twenty-five double oxbows containing 300 pieces
- 100 'pykeys'
- Fifty axes
- 200 shovels
- 200 spades
- Forty-two 'scoops'
- Twelve 'crowes' of iron

1 TNA, E 364/69, rot. Q dorse.

- One axle beam of iron
- Seven weights of lead
- 996 hammers of lead
- Seven big pieces of lead weighing 3,525lbs
- One pair of bellows for smiths
- 800 shafts called spear shafts
- 710 lance heads
- 391 chests for arrows and bows
- 1,204lbs of gunpowder
- 5,164lbs of saltpetre liquid not dry or pestled in powder
- 5,200lbs of sulphur rough not dry in powder
- 280lbs of charcoal
- One stock to hold saltpetre, sulphur and charcoal
- 100 long pavises
- Five short pavises
- Three cords of hemp called hankers
- Eighteen stocks of wood and iron not bound for guns called fowlers
- Forty-two empty vessels
- Eight hogsheads
- Fourteen barrels
- Six 'wassell' of metal
- Three saddles called 'lymour' saddles
- Three cruppers for 'lymour' saddles
- Three backs (dosers) for the same
- Two collars of rope for cart horses
- Four 'crebis of crunbz' for great ordnance

£2,072 4s. 8½d.

Appendix G

Extracts from the Accounts of John Hampton 1430–2 – Receipt of Equipment[1]

Received of Richard Bokeland treasurer and victualler of the town and marches of Calais for the use of the king by indenture between the same treasurer and master of the ordnance given 25 May in the eighth year of the reign (1430):

- Three large guns of iron, of which one called *Flouredelys* [Fleur de lis], one called *Newgate* and one called *Towerwharf of Cumbria*

Received from the stores of the king of William Appulby, master of the ordnance of the king in Normandy without indenture:

- One gun called *Coloffre*
- One gun of iron called a fowler fixed and bound in a stock of wood and iron
- Two chambers of iron for the same gun
- 1,672lbs of sulphur dry and in powder form
- 717lbs of gunpowder
- 100 bushels of willow charcoal
- Sixty-six empty vessels
- 228lbs of lead

Received of William Hampton master carpenter of the king at Rouen without indenture:

- Fourteen stocks of wood and iron not fixed or bounds for guns called cart guns
- Sixteen stakes of wood

[1] TNA, E 364/69, rot. Q dorse.

Received of the vicomte of Rouen without indenture:

- Sixteen carts of iron not bound to hold guns called cart guns

Received of John Clampard smith of the king at Rouen without indenture:

- Fourteen guns of iron called cart guns fixed and bound in stocks of wood and iron
- Twenty-eight chambers of iron for the same cart guns
- Fifteen chains of irons for the same cart guns
- Sixteen hooks of iron for the same
- Eighteen guns of iron called fowlers, of which four with bronze chambers and fourteen with iron chambers, bound and fixed in stocks of wood and iron
- Thirty-nine chambers for the same guns called fowlers
- Sixty-eight bolts of iron for loops for guns
- Three pairs of tongs for smiths
- Two 'toshornes' of iron
- Two 'bikornes' of iron
- Two mandrels of iron
- Two hooks of iron
- One chisel of iron
- One 'dribolt' of iron for smiths
- One 'cirall' of iron for moving piles of wood by water
- One pair of 'tonges' for moving stones from water and land
- Twenty pikes of iron

Received of William Warner gunpowder maker at London without indenture:

- 2,064lbs of saltpetre dry in powder form

Received of Robert (Westley?) at London without indenture:

- 4,648lbs of sulphur dry in powder form

Received of Robert (Westley?) at Rouen without indenture:

- 2,374lbs of saltpetre dry in powder form
- 7,925lbs of gunpowder

Received of the captain and other officers of the town of Harfleur:

- 1,664lbs saltpetre dry in powder form
- 1,933lbs saltpetre dry

Received of William Rothwell, esquire, by the hands of Richard Morkok by indenture given 24 January in the eighth year of the reign (1430):

- Forty oxen to move large guns of the king

Appendix H

Extracts from the Accounts of John Hampton 1430–2 – Expended or Given Away Equipment[1]

Expended in the time of the accounts:

- Eighteen guns of iron and bronze called fowlers of which four of bronze and fourteen of iron
- Eighteen stocks of wood for guns called fowlers
- Thirty-two chambers of iron for guns of iron called fowlers
- Eight chambers of bronze for guns of bronze called fowlers
- Two guns of bronze called fowlers fixed and positioned in stocks of wood and iron bound
- One chamber of bronze for the same
- 625 gunstones of diverse sorts
- 4,252lbs of iron not worked
- 2,128lbs of iron worked
- Fourteen stocks of wood and iron not bound for guns of iron called cart guns
- 6,113lbs of gunpowder
- 5,164lbs of saltpetre liquid not dry or powder
- 2,705lbs sulphur dry in powder
- Sixty-eight bushels of willow charcoal
- Six chaldrons and eight bushels of sea coal
- Five empty barrels
- 202 shafts called spear shafts
- Twenty pikes for scaling ladders
- Twenty-three pavises
- Twelve long pavises
- Two 'crowes' of iron
- One wedge of iron for guns called fowlers for shooting of the same
- Twenty-four shovels of iron

1 TNA, E 364/69, rot. Q dorse.

- Six 'wassels' of metal
- Eight hoggesheads pavises
- 231 hammers of lead
- Sixty-one spades
- Fourteen axes
- Nine chains of iron for moving ordnance
- Forty-four hooks with staples for the same of which twenty-two pairs of hooks and twenty-two staples
- Four long carts made for use in the country of Picardy iron bound
- Forty plates for neck halters
- 120 oxbows
- One pair of wheels called trokells,
- One saddle called 'lyino saddle'
- One cropper and back for the same 'lyino saddle'
- Six bolts of iron for 'loopes' for guns
- Two pairs of tongs of iron for smiths
- Two mandrels of iron

Given to William Appulby of the county of Lincoln, master of the ordnance of the king in Normandy by four indentures:

- Six large guns of iron, of which one called *Flourdelice*, one called *Newgate*, one called *Towerwharf*, one called *Henry*, one called *Crown* and one called *Hampton*
- One gun of iron called *Coloffr*
- Seven guns of iron called fowlers fixed and positioned in stocks of wood iron bound
- Fifteen chambers of iron for the same guns
- Fourteen guns of iron called cart guns fixed and positioned in stocks of wood iron bound
- Twenty-eight chambers of iron for the same cart guns
- Fifteen chains of iron for the same guns
- 765 stones for guns of diverse sorts
- 380lbs saltpetre dry in powder
- 3,539lbs sulphur dry in powder
- Seventy long pavises
- Three pieces of lead weighing 1,520lbs
- Twelve pieces of iron weighing 600lbs not worked
- Thirty-one bushels of carbon sea coal
- 120 bushels of willow charcoal
- Six barrels for the same charcoal
- Ten 'crowes' of iron

- 1,722lbs gunpowder
- Eight empty barrels for gunpowder
- 137 spades
- 170 shovels of iron
- Eighty-five pikes
- Thirty axes
- Sixty-three 'scoopers'
- 400 hammers of lead
- One pair of bellows for smiths
- One pair of tongs of iron for smiths
- One chisel of iron for smiths
- Two 'toshornes' for smiths
- Three cords of hemp called hansers for guns
- Twenty-one chains of iron for ordnance for horses
- Seventy-two bolts of iron for loups weighing 232lbs
- Two saddles called lyinosaddles
- Twenty cruppers for lyinosaddles
- Two backs for lyinosaddles
- Two collars of cords for cart horses
- Twenty-three halters of wood for horses
- One large cart iron bound with apparatus for moving guns
- Sixteen carts of iron not bound for carrying guns called cart guns
- Sixteen hooks of iron for the same carts
- Sixteen stakes of wood

Given to William Gloucester of the county of Gloucester in the duchy of Normandy, former master of the ordnance of the king in Normandy:

- One gun of bronze called fowler
- Five chambers of bronze for the same gun fixed and positioned in stocks of wood iron bound
- Two guns of iron called fowlers
- Two chambers of iron for the same fowlers fixed and positioned in stocks of wood iron bound
- Eighty stones for guns of diverse sorts
- 403lbs of gunpowder
- Two empty barrels for gunpowder
- 1,803lbs sulphur dry in powder
- 192 bushels of willow charcoal
- Twelve empty vessels for charcoal
- Twelve chalders and twenty-five bushels of seacole

- Four pieces of lead weighing 2,228lbs
- 225 lead hammers
- One axle beam of iron with one basin of laton and one axle of wood and cords of hemp for gunpowder and other things
- Seven pieces of lead weighing 101lbs
- 425lbs of iron not worked
- Two large carts iron bound for supporting and moving large guns
- Fifty-six collars of wood for horses
- 180 oxbows for the same halters
- Twenty chains of iron for oxen
- Eighteen staples for the same chains
- Eighteen staples and eighteen hooks
- Five chambers of iron for the guns called fowlers not hooped or whole (unfinished)
- One stock of wood with four external 'forauny' and four horse harnesses with hoops and axles of iron for saltpetre, sulphur and willow charcoal
- Two sieves
- Three 'scoopes'
- One 'ciront' of iron for moving piles out of water
- One pair of tonges of iron for moving stones out of water

Given to John, duke of Bedford, by the hands of John Falstolf, knight, of diverse things for the safekeeping of the castle and town of Caen by indenture made 16 December in the ninth year of the reign (1431) between John Falstolf and John Hampton:

- Two guns of iron called fowlers fixed and positioned in stocks of wood bound with iron
- Two guns of iron not fixed or positioned in wooden stocks iron bound
- Four chambers for the same four guns of which two chambers not fixed or positioned in stocks of wood iron bound
- 660lbs of gunpowder
- Two empty barrels for gunpowder
- Forty gunstones of diverse sort
- 200lbs dry sulphur in powder form
- 100 hammers of lead
- Twelve long pavises

Given to John, earl of Huntingdon, for the safekeeping of the castle and town of Gournay by indenture made 14 December anno 9 (1431) between John, earl of Huntingdon, and John Hampton:

- Three guns of iron called fowlers fixed and positioned in stocks of wood iron bound
- Six chambers of iron for the same guns
- One gun of bronze called fowler fixed and positioned in stock of wood iron bound
- Two chambers of metal for the same
- Two wedges of iron for the same guns called fowlers
- 660lbs gunpowder
- Two empty barrels for gunpowder

Given to John Feriby esquire of the county of Surrey for the safekeeping of the castle and town of Touques in Normandy by indenture made 5 December anno 9 (1431) between John Fereby and John Hampton:

- Three guns of iron called fowlers fixed and positioned in stocks of wood iron bound
- Six chambers of iron for the same gun
- One gun of bronze called fowler fixed and positioned in stock of wood iron bound
- Two chambers of bronze for the same
- Two wedges of iron for the same guns called fowlers
- 660lbs gunpowder
- Two empty barrels for gunpowder

Appendix I

Movements of Guns for the Blackheath Campaign of 1497[1]

Carriage of certain ordnance from the Tower of London to Woodstock
Paid by John Smyth, yeoman of the king's ordnance:

- The Tower of London to Uxbridge nine whole and eight half carriages – sixteen miles
- Uxbridge to Aylesbury nine whole and eight half carriages – eighteen miles
- Aylesbury to Stratton Audley nine whole and eight half carriages – ten miles
- Stratton Audley to Banbury nine whole and eight half carriages – ten miles
- Banbury to Dean nine whole and eight half carriages – twelve miles
- Dean to Woodstock nine whole and eight half carriages – (no miles given)

Total of sixty-six miles

Carriage of eight falcons from Woodstock to Blackheath
Paid by the hands of the master of the ordnance for the carriage of eight falcons:

- Woodstock to Oxford in two carriages – seven miles
- Oxford to Nettlebed two carriages – twelve miles
- Nettlebed to Maidenhead in two carriages – thirteen miles
- (Plus two carriages for artillery sent by the king from Wallingford to Henley and then to Maidenhead)
- Maidenhead to Hounslow Heath in five carriages – thirteen miles
- Carters paid for two days there to attend upon the ordnance with the lord chamberlain
- Hounslow Heath to Wimbledon in five carriages – twelve miles
- Wimbledon to St George's Field in five carriages – seven miles

1 TNA, E 36/8, ff. 76v–77v (note that these are the mileages recorded in the document, which do not correspond with the distances in modern miles).

- St George's Field to Streatham then St George's Field again in five carriages – eight miles
- St George's Field to Blackheath in five carriages – four miles

Total of seventy-six miles

Appendix J

Gun Categories

Bombard

Name variants: Large gun, bumbard, bumbarde, bunbard, bombacus, bumberd.[1]

Terminology: Guns of this type were initially described simply as being large. For example, an entry in the issue roll dated 16 January 1408 records a payment for making 'uni magni canonis'.[2] The term bombard began to be used around the mid-fifteenth century, with an entry for works carried out at Calais in October 1449 recording the making of loops for diverse 'bumberds'.[3] Its usage in Exchequer accounts increased in frequency during the reign of Edward IV. For example, an entry in the issue rolls in 1463 refers to the carriage of one 'bunbard de Werkworth' towards Bamburgh Castle.[4] Nevertheless, in some documents they continued to be described as large guns or simply guns, even in the late fifteenth century. For example, an inventory of Calais taken in 1476 includes 'j bumbard called Seyntpoule', but the same gun is consistently recorded in the accounts of the treasurer of Calais as merely a gun of iron for (firing) stones called 'Seynt Powle' from 1436 to 1488.[5]

Description: The largest and most prestigious type of gun used in the fifteenth century. These were used for knocking down the walls of enemy fortifications and were sometimes referred to by name in chronicles, such as during the siege

1 Dan Spencer, 'The Provision of Artillery for the 1428 Expedition to France', *Journal of Medieval Military History*, 13 (2015), 179–92 (p. 191); TNA, E 101/198/13, f. 63r; E 101/55/7; E 403/827A, m. 12; SC 6/780/2; E 101/194/2, m. 89.

2 TNA, E 403/594, m. 8.

3 TNA, E 101/194/2, m. 89.

4 TNA, E 403/827A, m. 12.

5 TNA, E 101/198/13, f. 63; E 364/72, rot. E; E 364/119/36, rot. D.

of Bamburgh Castle in 1464.[6] They were mostly constructed out of wrought iron but occasionally, particularly in the early part of the period, were made of bronze. Gunpowder specifically adapted for the use of bombards is first recorded as being purchased for the Calais garrison in 1475.[7] Bombards were mounted on wooden beds and were lifted by means of cranes referred to as 'gynnes' or engines onto large carts when they needed to be transported.[8] In 1418, it was specified that 12 large carts along with 320 oxen and horses were required for moving large guns. The same entry also records other equipment needed for these bombards, including 300 large pavises, 80 blocks of wood, 50 yokes of wood for oxen, 300 collars of leather with hooks of wood for oxen and horses, as well as 7,000 tampons.[9]

Date range: Large guns are first recorded in 1365 but by later standards these were rather small. It was not until the reign of Henry IV that large named guns, which correspond to the bombard class, are recorded. These were in existence by 1407 at the latest, during the siege of Aberystwyth, when two such guns, *Neelpot* and *Messenger*, were broken. Bombards were still in use in the early sixteenth century during the reign of Henry VIII.[10]

Weights: These were clearly very heavy guns, which increased in size throughout the course of the period.[11]

Date	Quantity	Material	Name of gun	Weight (lbs)	Average weight (lbs)
1400–15	1	Iron	*Fowler*	1,100	1,100
1400–15	1	Iron	*The George*	3,675	3,675
1400–15	1	Bronze	*Messenger*	4,480	4,480
1400–15	1	Bronze	*Gobette*	1,100	1,100
1400–15	1	Iron	*King's Daughter*	5,600	5,600
1428	3	Iron		15,500	5,167
1428	1	Iron		5,350	5,350

6 A. R. Myers, ed., *English Historical Documents, 1327–1485*, volume 4 (London: Eyre & Spottiswoode, 1969), pp. 289–90.

7 TNA, E 364/119/36, rot. B.

8 An example of an engine is recorded in John Sturgeon's inventory of 1475, TNA, E 101/55/7.

9 *CPR 1416–1422*, pp. 133–4.

10 Tout, 'Firearms in England in the Fourteenth Century', p. 692; TNA, E 364/49, rot. C; Howard L. Blackmore, *The Armouries of the Tower of London, I: Ordnance* (London: H. M. S. O., 1976), p. 220.

11 TNA, E 364/49, rot. C; E 101/51/27; E 364/69, rot. B dorse; E 364/90, rot. D.

Date	Quantity	Material	Name of gun	Weight (lbs)	Average weight (lbs)
1428	3	Iron		14,800	4,933
1430	1	Iron	*Henry*	6,780	6,780
1430	1	Iron	*Crown*	7,122	7,122
1432–4	1	Bronze		3,904	3,904
1453–6	1	Iron	*Goodgrace*	22,400	22,400
1453–6	1	Iron	*Henry*	19,040	19,040
1453–6	1	Iron	*Crown*	7,840	7,840

Dimensions: The most comprehensive set of dimensions available for bombards is recorded in the accounts of William Hickling.[12]

Type of gun	Name	Length (in feet)	Size of shot (in diameter)	Weight (in lbs)
Bombard	*Goodgrace*	12	22 inches	22,400
Bombard	*Henry*	11	19 inches	19,040
Bombard	*Crown*	9.5	15 inches	7,840

Ammunition: Bombards are only ever described as firing large gunstones. The size of shot is specified for seven bombards procured for the 1428 expedition to France: one is listed as firing gunstones fourteen inches in diameter, four of sixteen inches in diameter and two of eighteen inches in diameter. Some of these guns could fire even larger gunstones, with the same accounts recording seventy-four gunstones which were twenty-four inches in diameter.[13]

Individual names: The evidence indicates that it was standard practice for bombards to be given their own individual names. For instance, the accounts of John Hampton list the names of the six large guns procured for the royal expedition to France in 1430: *Henry, Hampton, Crown, Flouredelys, Newgate* and *Towerwharf of Cumbria*.[14]

12 Dan Spencer, 'The Lancastrian Armament Programme of the 1450s and the Development of Field Guns', *The Ricardian*, 25 (2015), 61–70 (p. 70).

13 Spencer, 'The Provision of Artillery for the 1428 Expedition to France', pp. 184–5.

14 TNA, E 364/69, rots. Q–Q dorse, R.

Bombardelle

Name variants: Bumbardett.[15]

Terminology: There was little variation in the spelling of this gun type. A typical example can be seen from a survey taken of the artillery at Guînes Castle in 1485, which records 'j bombardell with ij chamb', whereas another form of this spelling can be seen from a later set of accounts for 1492 which refers to two 'bumbardelles of brass'.[16]

Description: As noted by Howard Blackmore, this appears to be a smaller variant of a bombard, based on its name.

Date range: This type of gun was first recorded in an inventory of ordnance transferred by John Sturgeon, master of the ordnance, for the 1475 expedition to France.[17] By the 1490s, bombardelles appear to have assumed the place of bombards, as they were the largest guns recorded in the French and Scottish expeditions of 1492 and 1497. They were still in use in the early sixteenth century.[18]

Weights: Not recorded.

Dimensions: Two unnamed bombardelles are recorded in 1492 as each firing 273lbs of iron shot, which required 60lbs of gunpowder.[19] Further information on the dimensions of these guns is available from the accounts of Robert Clifford for the 1497 expedition to Scotland, which provides details on the weight of gunpowder and shot used to fire two stocked and carted bombardelles, along with other equipment allocated for their use, such as the number of wheels per carriage.[20] This information indicates that the size of these weapons varied considerably.

Name	Gunpowder (lbs)	Iron shot (lbs)	No. of wheels	Iron shot	Gunstones	Horse harnesses
Portcolies	60	220	4	65	30	26
Rose	60	120	4	67	36	26

15 TNA, E 101/55/7.
16 TNA, E 101/198/13, f. 2r; E 36/15, f. 8r.
17 TNA, E 101/55/7.
18 Blackmore, *The Armouries of the Tower of London*, p. 220.
19 TNA, E 36/15, f. 33v.
20 TNA, E 36/8, ff. 2r, 5v.

Ammunition: These guns appear to have primarily fired iron shot, but were also issued with gunstones for the 1497 expedition to Scotland.

Individual names: As these were large guns it is likely that it was standard practice for them to be given individual names. The two bombardelles recorded in the accounts for the 1497 expedition to Scotland were called *Portcolies* and *Rose*.

Cart gun

Name variants: Ribaudequin, ribaultquine, rebawkyn, rybaudekyn.[21]

Terminology: Guns of this type were initially described as *ribaudequins*, or variants of this term, which derived from the Dutch word *ribaudekijn*. For example, a warrant for issue dated 8 April 1443 mentions 'xx carts ribaultquines'.[22] Another variation can be seen from a later warrant for issue dated 23 June 1450, which lists various items of ordnance including 'v grete rebawkyns redy for the said feeld with x chambres'.[23] It has been argued by some scholars that the term *ribaudequin* refers to a carriage as opposed to a gun type.[24] However, it is clear from the evidence of the Exchequer that they were guns. This can be seen from the accounts of William Hickling, which refer to 'xxiiij canonis vocatur rybaudekyns cum xlviij cameris'.[25] Previous writers, such as Blackmore, have associated *ribaudequins* with organ guns, but this contention does not seem to be borne out by the evidence, as the former are only ever recorded as having two gun chambers, whereas the latter are described as having many organs.[26] The term *ribaudequin* fell out of usage in the reign of Edward IV, to be replaced with the English name cart gun. For example, a survey of ordnance at Rysbank Tower taken in October 1474 lists 'ij cart gonnes with v chambers and ij carts'.[27]

21 TNA, E 404/59, no. 162; E 404/66, no. 181; E 364/90, rot. D.

22 TNA, E 404/59, no. 162.

23 TNA, E 404/66, no. 181.

24 For example, see A. V. B. Norman, 'Notes on Some Early Representations of Guns and on Ribaudekins', *Journal of the Arms and Armour Society*, 8 (1975) 234–7.

25 TNA, E 364/90, rot. D.

26 Blackmore, *The Armouries of the Tower of London*, p. 237. The present author has also mistakenly associated *ribaudequins* with organ guns, as opposed to cart guns, in a previous publication: Spencer, 'The Lancastrian Armament Programme of the 1450s and the Development of Field Guns'. For a discussion on organ guns see below.

27 TNA, E 101/198/13, f. 69v.

Description: As the name indicates this was a type of gun that was mounted in a cart. The evidence suggests that they were very small in size.

Date range: This type of gun was first recorded in John Hampton's accounts for 1430–2 and remained in usage until the reign of Henry VII.[28]

Weights: In the accounts of William Hickling each of the twenty-five cart guns averaged 324lbs in weight.[29]

Dimensions: The most comprehensive set of dimensions available for cart guns is recorded in the accounts of William Hickling.[30]

Type of gun	Length (in feet)	Size of shot (in diameter)	No. of chambers	Length of chambers (in feet)	Average weight (lbs)
Cart gun	4	4 inches	2	1	338
Cart gun	4	4 inches	2	1	338
Cart gun	4	4 inches	2	1	338
Cart gun	4	4 inches	2	1	338
Cart gun	4	4 inches	2	1	338
Cart gun	4	4 inches	2	1	338
Cart gun	4	4 inches	2	1	338
Cart gun	4	4 inches	2	1	338
Cart gun	4	4 inches	2	1	338
Cart gun	4	4 inches	2	1	338
Cart gun	4	4 inches	2	1	338
Cart gun	4	4 inches	2	1	338
Cart gun	4	4 inches	2	1	338
Cart gun	4	4 inches	2	1	338
Cart gun	4	4 inches	2	1	338
Cart gun	4	4 inches	2	1	338
Cart gun	4	4 inches	2	1	338
Cart gun	4	4 inches	2	1	338

28 TNA, E 364/69, rots. Q–Q dorse; E 101/198/13, f. 57r.

29 Spencer, 'The Lancastrian Armament Programme of the 1450s and the Development of Field Guns', p. 70.

30 Ibid., p. 70.

Type of gun	Length (in feet)	Size of shot (in diameter)	No. of chambers	Length of chambers (in feet)	Average weight (lbs)
Cart gun	4	4 inches	2	1	338
Cart gun	4	4 inches	2	1	338
Cart gun	4	4 inches	2	1	338
Cart gun	3	4 inches	2	1	338
Cart gun	3	4 inches	2	1	338
Cart gun	3	4 inches	2	1	338

Ammunition: Cart guns were recorded as firing lead shot.[31]

Individual names: Not recorded.

Culverin

Name variants: Culverien, colovryue.[32]

Terminology: The name of this gun derives from its French equivalent *couleuvre*.[33] For example, a warrant for issue dated 23 June 1450 lists various items of ordnance purchased including 'j culverien'.[34] Another variant of this name can be seen from the accounts of the treasurer of Calais for 1467–8, which lists the purchase of 'vj gunnis de ferro vocatur colovryues'.[35]

Description: As stated by Howard Blackmore this was a type of gun which was long in proportion to its bore.[36] These varied in size, with the smaller models resembling handguns. This can be seen from an entry of purchases for Guînes Castle in 1472 which includes the mention of thirty-three culverins called hand culverins, together with nineteen pouches, stopping irons and cockers of the same, price of each one 4s., which totalled £19 16s.[37] Gunpowder specifically adapted for

31 SRO, SC 5/1/7, f. 16r.

32 TNA, E 404/66, no. 181; E 364/103, rot. E.

33 Emmanuel, de Crouy-Chanel, *Canons Médiévaux: Puissance du Feu* (Paris: Rempart, 2010), pp. 72–7.

34 TNA, E 404/66, no. 181.

35 TNA, E 364/103, rot. E.

36 Blackmore, *The Armouries of the Tower of London*, p. 224.

37 TNA, E 159/249, recorda, Hilary, rots. 12d–13.

the use of culverins is first recorded in the accounts for the treasurer of Calais in 1434–6.[38]

Date range: Culverins were first recorded in 1443, with the term still being used for guns in the sixteenth century.[39]

Weights: Not recorded.

Dimensions: Not recorded.

Ammunition: As small types of guns they primarily fired lead shot.[40]

Individual names: Not recorded.

Curtow

Name variants: Cortogh, cortegh, courtoux, curteux.[41]

Terminology: This type of gun originated in France with the name curtow deriving from the French *courteaux*.[42] They are first recorded in an inventory of ordnance transferred from the Tower of London to Calais in 1475, which included 'j curtowe of irenne and j carte of iiij wheles'.[43] Multiple variations of the spelling are recorded in English accounts. For example, an entry for 1489 includes a payment to Sir Richard Guildford for 'making divers guns of the king called curteux', whereas another set of accounts for the following year includes expenditure on 'ij cortoghs'.[44]

Description: This was a large type of gun that could fire both iron shot and gunstones and required numerous horses to transport. A reference from 1489 indicates that they were comparable to bombards, with a gunmaker paid for making 'bombardorum vocatur curtews'.[45]

38 TNA, E 364/72, rot. E.
39 *CPR 1441–1446*, p. 199; Blackmore, *The Armouries of the Tower of London*, p. 224.
40 For instance, see TNA, E 364/66, rot. C.
41 TNA, E 36/124, f.39r; E 405/78, mm. 1d, 2d.
42 Robert D. Smith and Kelly DeVries, *The Artillery of the Dukes of Burgundy 1363–1477* (Woodbridge: Boydell Press, 2005), p. 219.
43 TNA, E 101/55/7.
44 TNA, E 405/78, f. 3r; E 36/124, f. 39.
45 TNA, E 405/77, m. 6r.

Date range: This type of gun was first recorded in an inventory of ordnance transferred by John Sturgeon, master of the ordnance, for the 1475 expedition to France.[46] It was not until the Brittany campaigns of Henry VII, however, that they were used on a regular basis.[47] This type of gun was still in use in the early sixteenth century.[48]

Weights: Not recorded.

Dimensions: Two unnamed curtows are recorded in 1492 as each firing 100lbs of iron shot, which required 60lbs of gunpowder.[49] Further information on the dimensions of these guns is available from the accounts of Robert Clifford for the 1497 expedition to Scotland, which provides details on the weight of gunpowder and shot used to fire two stocked and carted curtows, along with other equipment allocated for their use, such as the number of wheels per carriage. This demonstrates that curtows varied in size, with the *Tower* described as being a 'great curtow'.[50]

Name	Gunpowder (lbs)	Iron shot	No. of wheels	Iron shot	Gunstones	Horse harnesses	Charging ladles	Rammers
Dragon	40	96	2	150	50	20	2	2
Tower	60	120	4	300		26		

Ammunition: These guns appear to have primarily fired iron shot, but were also issued with gunstones for the 1497 expedition to Scotland.

Individual names: As these were large guns it is likely that it was standard practice for them to be given individual names. The two curtows recorded in the accounts for the 1497 expedition to Scotland were called *Dragon* and *Tower*.

46 TNA, E 101/55/7.

47 For instance, see TNA, E 36/124, f. 37v.

48 Blackmore, *The Armouries of the Tower of London*, p. 225.

49 TNA, E 36/15, f. 33v.

50 TNA, E 36/8, ff. 1r, 5v.

Demi-curtow

Name variants: Curtew.[51]

Terminology: The terminology of this gun type is similar to that of curtows. For example, an inventory of 1492 includes the entry 'demi curtews of brass – xx'.[52]

Description: As the name indicates this gun was half the size of the bigger curtow, but it was still a large gun.

Date range: This term of gun is first recorded in the accounts for 1492 and was still in use in the early sixteenth century.[53]

Weights: Not recorded.

Dimensions: Demi-curtows are recorded in 1492 as each firing 40lbs of iron shot, which required 27lbs of gunpowder.[54] Further information on the dimensions of these guns is available from the accounts of Robert Clifford for the 1497 expedition to Scotland, which provides details on the weight of gunpowder and shot used to fire eleven stocked and carted demi-curtows, along with other equipment allocated for their use, such as the number of wheels per carriage.[55]

51 TNA, E 36/15, f. 8r. Also see variants for curtow above.

52 Ibid.

53 Blackmore, *The Armouries of the Tower of London*, p. 225.

54 TNA, E 36/15, f. 33v.

55 TNA, E 36/8, ff. 1r–6r.

Name	Gunpowder (lbs)	Iron shot (lbs)	No. of wheels	Iron shot	Gunstones	Horse harnesses	Charging ladles	Rammers
Carlisle	25	40	2	150	50	28		
Dover	25	40	2	150	50	28		
Berwick	25	40	2	150	50	20	2	2
Leopard	25	40	2	150	50	20		
Rysbank	25	40	2	150	50	20		
Calais	25	40	2	150	50	20		
Pontefract	25	40	2	150	50	20	2	2
Windsor	25	40	2	150	50	20	2	2
Gusynes	25	40	2	150	50	20		
Bamburgh	25	40	2	150	50	15	2	2
Hammes	25	40	2	150	50	15	2	2

Ammunition: These guns appear to have primarily fired iron shot, but were also issued with gunstones for the 1497 expedition to Scotland.

Individual names: As these were large guns it is likely that it was standard practice for them to be given individual names. The eleven demi-curtows recorded in the accounts for the 1497 expedition to Scotland were called *Carlisle, Dover, Berwick, Leopard, Rysbank, Calais, Pontefract, Windsor, Gusynes, Bamburgh* and *Hammes*.

Falcon

Name variants: Faulcon, faucon.[56]

Terminology: There was little variation in the terminology used for this gun type. A typical example can be seen from a set of accounts from 1497, which records a payment for the construction of 'x faucons of brass'.[57]

Description: This was a small type of gun which was primarily used on the battlefield, thereby replacing the role previously occupied by small serpentines. Falcons were used in large numbers in the French and Scottish campaigns of 1492 and 1497.

Date range: Falcons are first mentioned in 1489 and they continued to be used in the sixteenth century.[58]

Weights: Not recorded.

Dimensions: Falcons are recorded as being stocked and carted in the accounts for the 1497 expedition, with 1lb of gunpowder being needed to fire lead shot weighing 1lb.[59]

Ammunition: The only type of ammunition used for falcons was lead shot.

Individual names: Not recorded.

56 TNA, E 404/80, no. 68; E 36/8, f. 58r.

57 TNA, E 36/8, f. 58r.

58 TNA, E 404/80, no. 68; Blackmore, *The Armouries of the Tower of London*, p. 229.

59 TNA, E 36/8, f. 3r.

Fowler

Name variants: Fouler, veugler, foweller.[60]

Terminology: This type of gun is first recorded in a writ enrolled in the King's Remembrancer's Memoranda Roll for 1417–18, which lists munitions expended during the siege of Harfleur in 1415. These items included 'trois mills sys centz peres pur canons appellez grosses foulers et trois mill peres pur canons appellez petitz foullers'.[61] The term fowler was also used in accounts written in Latin or English. For example, the accounts of John Hampton for 1430–2 list 'xvj canonis de ferro vocatur fowellers'.[62] Similarly, an inventory of ordnance transferred from the Tower of London to Calais in 1475 includes a gun 'called the fowler of the towre'.[63] By contrast, the equivalent term typically used in French and Burgundian documents, *veuglaire*, is rarely recorded in English accounts. One of the few examples of this usage can be seen from the accounts of William Hickling, which records 'iiij canonis vocatur veuglers'.[64]

Description: This was the single most popular type of gun recorded in English accounts in the fifteenth century. These weapons varied considerably in size, with the earliest known mention in 1415 referring to both small and large fowlers.[65] Fowlers were used in small numbers on royal ships in the first half of the fifteenth century, but appear to have been most effective in offensive or defensive siege warfare.[66] As with serpentines, these guns were regularly described as having multiple removable powder chambers, which allowed them to be reloaded and fired more quickly.[67] Fowlers were often described as being stocked or mounted on wooden trestles, although some were listed as having their own carts in the late fifteenth century.[68]

Date range: Fowlers are first mentioned in 1415 and they continued to be used in the sixteenth century.[69]

60 Spencer, 'The Provision of Artillery for the 1428 Expedition to France', p. 191; TNA, E 364/90, rot. D; E 364/69, rot. Q dorse.

61 TNA, E 159/194, brevia directa baronibus, Michaelmas, rot. 6.

62 TNA, E 364/69, rot. Q dorse.

63 TNA, E 101/55/7.

64 TNA, E 364/90, rot. D.

65 TNA, E 159/194, brevia directa baronibus, Michaelmas, rot. 6.

66 For the use of fowlers on royal ships, see Chapter 3.

67 For an example of this for 1472, see TNA, E 159/249, recorda, Hilary, rots. 12d–13.

68 For examples see TNA, E 101/198/13.

69 TNA, E 159/194, brevia directa baronibus, Michaelmas, rot. 6; Blackmore, *The Armouries of the Tower of London*, pp. 230–1.

Weights: The weights of this type of gun varied considerably between the smaller and larger fowlers.[70]

Date	Quantity	Material	Weight (lbs)	Average weight (lbs)
1428	2	Bronze	168	84
1453–6	4	Iron	7,840	1,960
1468–9	1	Iron	112	112
1468–9	2	Iron	717	358
1472	1	Iron	233	233
1472	1	Iron	340	340
1472	1	Iron	336	336
1472	1	Iron	560	560
1472	1	Iron	450	450
1472	1	Iron	440	440
1472	1	Iron	3,100	3,100
1472	1	Iron	1,600	1,600
1472	1	Iron	1,900	1,900
1472	1	Iron	1,100	1,100
1472	1	Iron	600	600
1472–3	14	Iron	5,707	408
1477–8	9	Iron	2,100	233
1479–80	3	Iron	697	232

Dimensions:

Some of the earliest fowlers were clearly very small in size, as can be seen from the accounts for the 1428 expedition to France, which records the dimensions of two fowlers.[71]

Material	Weight of shot (lbs)	Length (feet)	Individual weight (lbs)
Bronze	2	1.5	84

70 TNA, E 101/51/27; E 364/90, rot. D; E 364/103, rot. G; E 159/249, recorda, Hilary, rots. 12d–13; E 364/106, rot. I; E 101/55/4.

71 Spencer, 'The Provision of Artillery for the 1428 Expedition to France', p. 191.

The most comprehensive set of dimensions available for fowlers is recorded in the accounts of William Hickling.[72]

Type of gun	Length (in feet)	Size of shot (in diameter)	No. of chambers	Total weight of gun type
Fowler	8	10 inches	2	1,960
Fowler	8	10 inches	2	1,960
Fowler	8	9 inches	2	1,960
Fowler	8	9 inches	2	1,960

Some fowlers were even larger than this, as can be seen in an inventory of guns taken at Caister Castle in 1462, which records that three fowlers shot gunstones of twelve inches in diameter.[73]

Ammunition: The only recorded type of ammunition used by fowlers was gunstones.

Individual names: The larger forms of fowlers were occasionally given their own names, for example the *Gloucester*, which was acquired by the Calais garrison in 1436.[74]

Hackbut

Name variants: Hake gun, hagbusshe, hakbusse.[75]

Terminology: A variety of terms were used to describe this type of gun. This can be seen from a set of accounts for 1492 which alternates between describing these guns as 'haggbushses', 'hakebusshes' and 'hakbusshes'.[76]

72 Spencer, 'The Lancastrian Armament Programme of the 1450s and the Development of Field Guns', p. 70.

73 Note that the term used in the inventory is 'cumpas', Norman Davis, ed., *Paston Letters and Papers of the Fifteenth Century*, part 1 (Oxford: Oxford University Press, 2004), pp. 112–13.

74 TNA, E 364/87, rot. H.

75 TNA, E 101/198/13, f. 8v; PSO 1/49, no. 2526A; E 36/8, f. 10r.

76 TNA, E 36/15, ff. 8r, 25r, 33v.

Description: As noted by Bert Hall this refers to a hook gun, which could be mounted on fortifications or a trestle.[77] Hackbuts were almost certainly heavier than other handguns and are often described as having their own removable powder chambers. This meant that they needed to be placed on trestles or wooden stocks, with pavises provided for protecting their operators.[78] At least some of these weapons were imported, with bronze hackbuts purchased from Brabant in the late 1470s.[79] Interesting information is provided in a description of equipment sent to Guînes Castle in 1481, with one iron hackbut described as having 'dagons and chambers of hym selfe and j smale gon with a bit and a chamber of hym selfe in the same dragons hede stokked … viij hake gonnes of yron with viij chambers of theym self and viij stampers redy stokked'.[80] Gunpowder specifically adapted for the use of hackbuts is first recorded as being purchased for the Calais garrison in 1475.[81]

Date range: This type of gun is first recorded in the accounts of the Calais garrison in 1469.[82] It is unclear when the term went out of usage.

Weights: In William Rosse's accounts for 1476–7, thirty bronze hackbuts each averaged 35lbs in weight, whereas eighty-six bronze hackbuts each averaged 29lbs in weight in 1479–80.[83]

Dimensions: Not recorded.

Ammunition: The only type of ammunition used for hackbuts was lead shot.

Individual names: Not recorded.

77 Bert S. Hall, *Weapons and Warfare in Renaissance Europe* (London: The John Hopkins Press Ltd, 1997), p. 176.

78 For an example of this see TNA, E 36/8, f. 78r.

79 TNA, E 101/55/4, f. 12r.

80 TNA, E 101/198/13, f. 8v.

81 TNA, E 364/119/36, rot. B.

82 TNA, E 364/103, rot. G dorse.

83 TNA, E 101/55/4, ff. 12r–12v..

Handgun

Name variants: Canonis manualis.[84]

Terminology: There were two main variations in the terminology used for this gun type. The first can be seen from a writ dated 7 November 1388 which refers to 'iij canones paruos vocatos handgunnes', whereas a later document from 1428 instead refers to 'xvj aliis parvis canonis manualis ad iactand pelletts de plumbo'.[85]

Description: The earliest handguns were small guns which could be moved easily by hand.[86] From the 1450s onwards, the term handgun was used to refer to a gun which was later called an 'arquebus', a predecessor to the musket. Large numbers of handguns were purchased for the use of the Calais garrison in the 1470s and 1480s. Handguns were relatively light and easy to transport, which meant that they were useful on expeditions but could also be kept in storage in the fortifications of Calais ready for the use of the garrison.[87] The handgun remained a niche weapon in the fifteenth century, however, which meant that it did not supplant the longbow.

Date range: The earliest mention of a handgun occurs in 1388, but in the second half of the fifteenth century the term was mainly used to describe a new type of weapon, which corresponds to an arquebus.[88] The term handgun is still used to describe firearms today.

Weights: Not recorded.

Dimensions: Not recorded.

Ammunition: The only type of ammunition used for handguns was lead shot.[89]

Date	Quantity	Type	Individual weight of shot (lbs)
1472–5	4000	Lead	0.17

84 Spencer, 'The Provision of Artillery for the 1428 Expedition to France', p. 191; TNA, E 101/51/27.

85 Tout, 'Firearms in England in the Fourteenth Century', p. 699. TNA, E 101/51/27.

86 For a discussion of early handguns see Tout, 'Firearms in England in the Fourteenth Century', p. 684.

87 For a discussion of this see Chapter 4.

88 Tout, 'Firearms in England in the Fourteenth Century', p. 699; for the emergence of the arquebus in the 1450s, see Hall, *Weapons and Warfare in Renaissance Europe*, p. 95.

89 TNA, E 101/55/4, f. 30v.

Individual names: Not recorded.

Murderer

Name variants: Not recorded.

Terminology: There was little variation in the spelling of this gun type. A typical example can be seen from the accounts of the Corporation of Dartmouth in 1487 which record the purchase of 'ij grete murderers for the castell'.[90]

Description: Little is known about this type of gun, although it appears to have been large in size and used for the defence of fortifications, such as Dartmouth Castle in the late 1480s.[91]

Date range: Murderers are first recorded in the Sandwich inventory of 1482–3 and were still in use in the sixteenth century.[92]

Weights: Not recorded.

Dimensions: Not recorded.

Ammunition: This was clearly a large type of gun as indicated by the weight of shot for ammunition purchased for large murderers in Dartmouth Castle.[93]

Date	Quantity	Type	Individual weight of shot (lbs)
1487–8	50	Lead and iron	20

Individual names: Not recorded.

90 DHC, DD 61199a.
91 Ibid.
92 Blackmore, *The Armouries of the Tower of London*, p. 236.
93 DHC, DD 61199a.

Organ guns

Name variants: Not recorded.

Terminology: There was little variation in the spelling of this gun type. A typical example can be seen from the accounts of the treasurer of Calais for 1466, which records 'j gunne de ere vocatur organs cum xxxix organs'.[94] Similarly, the steward book of Southampton for 1450 records payments for repairs to 'the carte of the organ gunne'.[95]

Description: As noted by Howard Blackmore, this type of gun was given its name because it resembled the musical instrument, consisting of a cluster of small barrels.[96] Organ guns were mounted in carts and could potentially fire many projectiles, as is indicated by the accounts of the treasurer of Calais for 1466–8, which records a bronze gun with as many as thirty-nine 'organs'.[97]

Date range: This type of gun was first recorded in 1450 and was still in use in the sixteenth century.[98]

Weights: Not recorded.

Dimensions: Not recorded.

Ammunition: Organ guns are only recorded as firing lead shot.[99]

Individual names: Not recorded.

Pellet gun

Name variants: Pelotgunne.[100]

Terminology: There was little variation in the spelling of this gun type. For example, the accounts of John Norbury, keeper of the privy wardrobe, for

94 TNA, E 364/101, rot. I dorse.
95 SRO, SC 5/1/17, f. 16r.
96 Blackmore, *The Armouries of the Tower of London*, pp. 236–7.
97 TNA, E 364/101, rot. I dorse; E 364/103, rot. G.
98 SRO, SC 5/1/17, f. 16r; Blackmore, *The Armouries of the Tower of London*, pp. 236–7.
99 SRO, SC 5/1/7, f. 16r.
100 TNA, E 364/40, rot. A.

1403–5 records 'ij pelotgunnes.'[101] Whereas a later entry, from the accounts of the chamberlain of south Wales for 1435–6, refers to 'ij pellet gunnes.'[102]

Description: Pellet guns were most commonly listed as being present in castle or town inventories. For instance, in 1434, two pellet guns were recorded in the storehouse at Southampton.[103] The popularity of this type of gun decreased markedly in the 1450s, as can be seen from the accounts of the treasurer of Calais, most likely as it was superseded by the serpentine, with no new pellet guns purchased after 1451.[104]

Date range: Guns that fired pellets are the earliest type of gun recorded in the accounts of the privy wardrobe, with the name pellet gun first recorded in John Norbury's accounts for 1403–5.[105] This type of gun was listed as being present in the inventories of the treasurer of Calais from 1375 until the last surviving account in 1488.[106] It is unclear when this term went out of usage.

Weights: The privy wardrobe accounts for 1382–8 indicate that these were light guns, with the nine bronze pellet guns each recorded as weighing 42½lbs on average.[107]

Dimensions: Not recorded.

Ammunition: As the name indicates this type of gun fired lead shot.

Individual names: Not recorded.

101 Ibid.

102 TNA, E 364/73, rot. N.

103 Harry W. Gidden, ed., *The Steward's Books of Southampton, from 1428*, volume 1 (Southampton: Cox & Sharland, 1935), pp. 90–3.

104 See the Calais accounts, such as TNA, E 364/87, rot. H, E 364/92, rot. C dorse.

105 TNA, E 364/40, rot. A.

106 TNA, E 364/14, rot. I; E 364/119/36, rot. D.

107 Tout, 'Firearms in England in the Fourteenth Century', pp. 697–8.

Pot gun

Name variants: Potgonne.[108]

Terminology: There was little variation in the spelling of this gun type. A typical example can be seen from the inventory of artillery transferred from the Tower of London to Calais in 1475, which includes 'ij grete potgonnes of bras'.[109]

Description: This was a large type of gun, which according to Howard Blackmore was 'either a gun with a separate chamber, or a mortar'.[110] The limited available evidence means that it is difficult to determine its form or function. In an inventory of 1475, three out of four pot guns were listed as being made of bronze, whereas one was constructed of iron.[111]

Date range: This type of gun was first recorded in an inventory of ordnance transferred by John Sturgeon, master of the ordnance, for the 1475 expedition to France.[112] They were still in in use in the sixteenth century.[113]

Weights: In the inventory of guns purchased for Guînes Castle by Richard Whetehill in 1472, one pot gun of iron was described as being 2,000lbs in weight.[114]

Dimensions: Not recorded.

Ammunition: The only type of ammunition used for pot guns was gunstones.

Individual names: Not recorded.

108 TNA, E 101/55/7.
109 Ibid.
110 Blackmore, *The Armouries of the Tower of London*, p. 239.
111 TNA, E 101/55/7.
112 Ibid.
113 Blackmore, *The Armouries of the Tower of London*, p. 239.
114 TNA, E 159/249, recorda, Hilary, rot. 12d.

Quarrel gun

Name variants: Not recorded.

Terminology: There was little variety in the limited number of references to this type of gun, with the issue roll for 1400 including a payment for 'xxiiij quarrel gunnes.'[115]

Description: Very little is known about this type of gun, which only briefly appears in Exchequer accounts early in the fifteenth century. In 1400, twenty-four quarrel guns were recorded as being purchased for the use of the king at 7s. the piece.[116]

Date range: These guns appear to have been in existence from the fourteenth century until the beginning of the fifteenth century.

Weights: Not recorded.

Dimensions: Not recorded.

Ammunition: As the name indicates these guns were intended to fire quarrel bolts.

Individual names: Not recorded.

Serpentine

Name variants: Sipontyne, sipent, serpent.[117]

Terminology: There was little variety in the terminology used with this gun type, with a typical example from the accounts of William Hickling referring to 'iiij canonis vocatur serpentynes.'[118]

Description: As noted by Howard Blackmore this was a long gun relative to its calibre which varied considerably in size.[119] Serpentines were first recorded in

115 TNA, E 403/567, m. 1.
116 Ibid.
117 TNA, E 101/198/13, f. 2r.
118 TNA, E 364/90, rot. D.
119 Blackmore, *The Armouries of the Tower of London*, p. 242.

French inventories from 1442, which may have given the French a technological advantage during their conquest of Normandy in 1450.[120] The introduction of serpentines in the late 1440s to early 1450s transformed English warfare, as they were soon used regularly on the battlefield. This was because the long barrel of this weapon meant that it had greater range and accuracy, which made it more effective in targeting enemy personnel. Large numbers of serpentines were acquired by the crown from the 1450s until the 1480s, which made it the most popular type of weapon, along with fowlers, for this period.[121] As with fowlers, these guns were regularly described as having multiple removable powder chambers, which allowed them to be reloaded and fired more quickly.[122] Serpentines were often described as being stocked or mounted on wooden trestles, although the mobility of this gun meant that many were also listed as possessing their own carts.[123]

The surviving views of the fortifications of Calais for the 1470s and 1480s make clear that serpentines were an important weapon both offensively and defensively in the second half of the reign of Edward IV.[124] They were also the main type of gun used on royal warships in the 1470s, 1480s and 1490s.[125] Gunpowder specifically adapted for the use of serpentines is first recorded in the accounts for the treasurer of Calais in 1461–2.[126] An unusual detail is provided in the accounts of purchases made by William Rosse for the 1475 expedition to France, with the mention of a long red serpentine of iron mounted on a cart with four wheels, which was acquired from a merchant of Brussels.[127]

Date range: Serpentines are first recorded in an inventory of 1448 and they were still in use in the sixteenth century.[128]

Weights: The greatest variation in weight is recorded for this type of gun, which reflected the many different ways that serpentines were used.[129]

120 See Chapter 1 for a full discussion of this.

121 For an example of this for 1472 see TNA, E 159/249, recorda, Hilary, rots. 12d–13.

122 For examples of this see ibid.

123 For examples see TNA, E 101/198/13.

124 Ibid.

125 For instance, see TNA, E 405/57, m. 7d.

126 TNA, E 364/96, rot. C.

127 TNA, E 101/55/5, f. 6.

128 MCO, FP 43, f. 13r; Blackmore, *The Armouries of the Tower of London*, p. 242.

129 TNA, E 364/90, rot. D; E 404/71/1, no. 60; E 364/103, rots. E, G; E 159/249, recorda, Hilary, rots. 12d–13; E 364/106, rot. I; E 101/55/4.

Date	Quantity	Material	Weight (lbs)	Average weight (lbs)
1453–6	4	Iron	840	210
1457	26	Iron	5,200	200
1467–8	1	Bronze	700	700
1467–8	1	Iron	858	858
1468–9	1	Iron	907	907
1468–9	1	Iron	230	230
1468–9	1	Iron	378	378
1468–9	1	Iron	529	529
1472	1	Iron	208	208
1472	1	Iron	585	585
1472	2	Iron	485	242.5
1472	3	Iron	340	113
1472	3	Iron	200	67
1472	1	Iron	280	280
1472	1	Iron	180	180
1472	1	Iron	4,200	4,200
1472	1	Iron	2,100	2,100
1472	1	Iron	1,400	1,400
1472	1	Iron	750	750
1472	1	Iron	750	750
1472–3	5	Iron	2078	416
1476–7	15	Bronze	4,225	282
1477–8	20	Iron	8,204	410
1478–9	4	Bronze	2,260	565
1480–1	4	Iron	2,946	736.5
1481–2	9	Iron	8,896	988
1481–2	1	Iron	1,725	1,725
1483–4	15	Iron	8,120	541

Dimensions: The most comprehensive set of dimensions available for serpentines is recorded in the accounts of William Hickling.[130]

Type of gun	Length (in feet)	Weight of shot (in lbs)	No. of chambers	Average weight
Serpentine	8.5	3.5	2	210
Serpentine	8.5	3.5	2	210
Serpentine	7.5	2.5	2	210
Serpentine	7.5	2.5	2	210

Some of the larger serpentines also used gunstones, as can be seen from an inventory of guns at Caister Castle in 1462, which records that one serpentine fired gunstones of ten inches in diameter and the other fired gunstones of seven inches in diameter.[131] Further information is also available from the accounts of Robert Clifford for the 1497 expedition to Scotland, which provides information on the weight of gunpowder and shot used to fire six stocked and carted serpentines, along with other equipment allocated for their use, such as the number of wheels per carriage. This demonstrates that serpentines varied considerably in size.[132]

Gunpowder (lbs)	Iron or lead shot (lbs)	No. of wheels	Iron shot	Horse harnesses	Charging ladles	Rammers
20	20	2	312	20	2	2
6	6	2				
6	6	2				
6	6	2	466	10		
6	6	2				
20	20	2	200	20		3

Ammunition: The main form of ammunition for serpentines in the fifteenth century was lead shot, although larger pieces were also capable of firing gunstones. By the mid-1480s, iron shot was also being increasingly used alongside lead shot, as can be seen in an inventory for Calais in 1485 which records '368 pellets of iron for large serpentines.'[133]

130 Spencer, 'The Lancastrian Armament Programme of the 1450s and the Development of Field Guns', p. 70.

131 Note that the term used in the inventory is 'cumpas', Davis, *Paston Letters and Papers of the Fifteenth Century*, pp. 112–13.

132 TNA, E 36/8, ff. 1r, 3r–v, 4v, 5v.

133 TNA, E 101/55/4, f. 30v; E 364/119/36, rot. B.

Date	Quantity	Type	Individual weight of shot (lbs)
1472–5	2,000	Lead	1.5
1485	6,000	Lead	1.1

Individual names: Serpentines were very rarely given their own individual names, although one was recorded as being called the *Greyhound* in 1497.[134]

Trestle gun

Name variants: Trestill gonne.[135]

Terminology: There was little variation in the spelling of this gun type. A typical example can be seen from a survey of artillery taken at Guînes Castle in 1481, which refers to two 'trestyll gonnes of yron.'[136]

Description: These were small guns placed on wooden trestles in the walls and towers of fortifications, as can be seen by the treasurer accounts of Sandwich for 1482–3, which record a payment to two men for 'trestelyng of the gones in the towre at davies gate.'[137] This was not a separate form of gun itself, but merely described how certain pieces were deployed, as it made clear in a view of Guînes Castle in 1483: 'j serpent with iij chambers redy trestylled … j fowler with iij chambers redy trestylled.'[138]

Date range: Trestle guns are mentioned in the inventory of Southampton's ordnance in 1467–8, but it is likely that they were used prior to that date.[139] It is unclear when this term went out of usage.

Weights: Not recorded.

Dimensions: Not recorded.

Ammunition: As trestle guns were mostly fowlers and serpentines mounted on trestles, this meant that they fired a mixture of gunstones and lead shot.

134 TNA, E 36/8, f. 5v.
135 TNA, E 101/198/13, f. 3r.
136 TNA, E 101/185/13, f. 85v.
137 KHLC, SA/FAT8.
138 TNA, E 101/198/13, f. 86v.
139 SRO, SC 5/1/7, f. 16v.

Individual names: A view of Calais taken in 1476 records 'j trestill gun of brass called Analide'.[140] This appears to correspond to a bronze gun captured by the Calais garrison in 1436, which was referred to in the accounts of the treasurer of Calais as being called the *Dame Anable*.[141]

140 TNA, E 101/198/13, f. 64r.
141 TNA, E 364/72, rot. E.

Bibliography

MANUSCRIPT COLLECTIONS

Bibliothèque Nationale de France
Français 26050

British Library, London
Additional Manuscripts
Additional Charters
Egerton Manuscripts
Royal Manuscripts

Canterbury Cathedral Archive
Treasurers' Account Rolls and Rentals

Devon Heritage Centre
Dartmouth Corporation Records
Exeter Receivers' Rolls

Durham University Library
Special Collections, Miscellanea on Accounts

East Sussex Record Office
Archive of Rye Corporation, Account Books

Hampshire Record Office
Winchester City Archives, City Account Rolls

Hull History Centre
Records of the Corporation of Kingston upon Hull, Bench Books

Kent History and Library Centre
Lydd Borough and Town Council, Chamberlains' Accounts
New Romney Borough, Chamberlains' Accounts
Sandwich Borough, Treasurers' Accounts
Sandwich Borough, Year Books

King's Lynn Borough Archives
Chamberlains' Account Rolls

London Metropolitan Archives
London Journal Books

Magdalen College Oxford
Fastolf Papers

The National Archives, Kew
C 47 – Chancery Miscellanea
C 76 – Chancery, Treaty Rolls
C 81 – Chancery, Warrants for the Great Seal
DL 37 – Duchy of Lancaster and Palatinate of Lancaster, Chanceries, Enrolments
DL 42 – Duchy of Lancaster, Cartularies, Enrolments, Surveys and other Miscellaneous Books
E 28 – Exchequer, Treasury of the Receipt, Council and Privy Seal Records
E 36 – Exchequer, Treasury of the Receipt, Miscellaneous Books
E 101 – King's Remembrancer, Accounts Various
E 159 – King's Remembrancer, Memoranda Rolls
E 315 – Court of Augmentations and Predecessors and Successors, Miscellaneous Books
E 364 – Exchequer, Pipe Office, Foreign Account Rolls
E 403 – Exchequer of Receipt, Issue Rolls
E 404 – Exchequer of Receipt, Warrants for Issue
E 405 – Exchequer of Receipt, Tellers' Rolls
PROB 11 – Prerogative Court of Canterbury, Will Registers
PSO 1 – Privy Seal Office, Signet and other Warrants for the Privy Seal, Series I
SC 6 – Special Collections, Ministers Accounts
SC 8 – Special Collections, Ancient Petitions

Norfolk Record Office
City of Norwich Records, Chamberlains' Accounts

Plymouth and West Devon Record Office

Plymouth Borough Records, Receivers' Accounts

Southampton Record Office
Stewards' Books of Southampton

Westminster Abbey Library and Muniment Room
MS 12240

Wiltshire and Salisbury Record Office
City Ledger Book B

PRINTED PRIMARY SOURCES

Anderson, R. C., ed., *Letters from the Fifteenth and Sixteenth Centuries: From the Archives of Southampton* (Southampton: Cox & Sharland, 1921)

Anderson, R. C., ed., *The Assize of Bread Book, 1477–1517, Publications of the Southampton Record Society* (Southampton: Cox & Sharland, 1923)

Arthur, Louis, ed., *Le Complot Breton de M.CCCC.XCII: Documents Inédits Publiés avec Notes et Introduction par A. de La Borderie* (Nantes: La Société Des Bibliophiles Bretons, 1884)

Attreed, Lorraine C., ed., *The York House Books: 1461–1490*, 2 vols. (Gloucester: Alan Sutton for Richard III & Yorkist History Trust, 1991)

Bain, Joseph, ed., *Calendar of Documents Relating to Scotland, Volume 4. A.D. 1357–1509. Addenda – 1221–1435* (Edinburgh: H. M. General Register House, 1888)

Bruce, John, ed., *Historie of the Arrivall of Edward IV: In England and the Finall Recouerye of his Kingdomes From Henry VI A.D. M.CCCC.LXXI* (London: Printed for the Camden Society by J. B. Nichols and Son, 1838)

Butler, Cheryl, ed., *The Book of Fines: The Annual Accounts of the Mayors of Southampton. Vol.1, 1488–1540* (Southampton: University of Southampton, 2007)

Calendar of the Close Rolls (H. M. S. O., 1900–63)

Calendar of the Patent Rolls (H. M. S. O., 1900–)

Carr, David R., ed., *The First General Entry Book of The City of Salisbury 1387–1452* (Trowbridge: Wiltshire Record Society, 2001)

Cooper, Charles Purton, ed., *Appendices to a Report on Rymer's Fœdera* (Public Record Office, 1869)

Crawford, Anne, ed., *The Household Books of John Howard, Duke of Norfolk, 1462–1471, 1481–1483* (Stroud: Sutton for Richard III & Yorkist History Trust, 1992)

Dallaway, James, ed., *Antiquities of Bristow in the Middle Centuries; Including the Topography by William Wycestre, and the Life of William Canynges* (Bristol: Mirror Office, 1834)

Davies, John Sylvester, ed., *An English Chronicle of the Reigns of Richard II, Henry IV, Henry V, and Henry VI, Written Before the Year 1471* (London: Camden Society, 1856)

Davis, Norman, ed., *Paston Letters and Papers of the Fifteenth Century*, part 1 (Oxford: Oxford University Press, 2004)

Devon, Frederick, ed., *Issues of the Exchequer: Being a Collection of Payments Made out of His Majesty's Revenue, From King Henry III to King Henry VI Inclusive: With an Appendix / Extracted and Translated from the Original Rolls of the Ancient Pell Office* (London: J. Murray, 1837)

Dobson, R. B., ed., *York City Chamberlain's Account Rolls 1396–1500* (Durham: Surtees Society, 1980)

Dormer, Mary, ed., *The Coventry Leet Book, or, Mayor's Register, Containing the Records of the City Court Leet or View of Frankpledge, A.D. 1420–1555, with Divers Other Matters* (Oxford: Oxford University Press, 1907–13)

Doüet-d'Arcq, L., ed., *La Chronique d'Enguerran de Monstrelet, en Deux Livres avec Pièces Justificatives, 1400–1444*, 6 vols. (Paris: Société de l'Histoire de France, 1857–62)

Fenn, John, ed., *Original Letters, Written During the Reigns of Henry VI, Edward IV, and Richard III* (London: G. G. J. and J. Robinson, 1787)

Fifth Report of the Royal Commission on Historical Manuscripts (H. M. S. O., 1877–8)

Fourth Report of the Royal Commission on Historical Manuscripts (H. M. S. O., 1874)

Gairdner, James, ed., *Letters and Papers Illustrative of the Reigns of Richard III and Henry VII*, 2 vols. (London: Longman, Green, Longman, and Roberts, 1861–3)

Gairdner, James, ed., *The Paston Letters: A.D. 1422–1509*, 6 vols. (London: Chatto & Windus, 1904)

Gidden, Harry W., ed., *The Book of Remembrance of Southampton, Volume 1* (Southampton: Cox & Sharland, 1927)

Gidden, Harry W., ed., *The Book of Remembrance of Southampton, Volume 2* (Southampton: Cox & Sharland, 1928)

Gidden, Harry W., ed., *The Steward's Books of Southampton, from 1428*, 2 vols. (Southampton: Cox & Sharland, 1935)

Given-Wilson, Chris, ed., *The Parliament Rolls of Medieval England 1275–1504*, 16 vols. (Woodbridge: Boydell Press, 2005)

Halliwell, James Orchard, ed., *A Chronicle of the First Thirteen Years of the Reign of King Edward the Fourth by John Warkworth, D. D. Master of St Peter's College, Cambridge* (London: Camden Society, 1839)

Harriss, G. L., and M. A. Harriss, eds., 'John Benet's Chronicle for the Years 1400 to 1462', in *Camden Miscellany XXIV*, Camden 4th ser., vol. 9 (1972)

Horrox, Rosemary, and P. W. Hammond, eds., *British Library Harleian Manuscript 433. Vol. 1, Register of Grants for the Reigns of Edward V and Richard III* (Gloucester: A. Sutton for the Richard III Society, 1979)

Horrox, Rosemary and P. W. Hammond, eds., *British Library Harleian Manuscript 433. Vol. 2, Second Register of Richard III* (Gloucester: A. Sutton for the Richard III Society, 1980)

Horrox, Rosemary, ed., 'Financial Memorandum of the Reign of Edward V: Longleat Miscellaneous Manuscript Book II', *Camden Miscellany*, XXIX. Camden 4th ser. XXXIV (London: Offices of the Royal Historical Society, 1987)

Hudson, William, ed., *The Records of the City of Norwich*, 2 vols. (Norwich: Corporation of the City of Norwich, 1898)

Isaacson, R. F., ed., *The Red Register of King's Lynn*, 2 vols. (King's Lynn: Thew & Son, 1919–22)

Kirby, J. L., ed., *Calendar of Signet Letters of Henry IV and Henry V (1399–1422)* (London: H. M. S. O., 1978)

Knighton, C. S., and D. M. Loades, eds., *The Anthony Roll of Henry VIII's Navy* (Aldershot: Ashgate, 2000)

Livingston, Michael, and Kelly DeVries, eds., *The Battle of Crecy: A Casebook* (Liverpool: Liverpool University Press, 2015)

Lyte, H. C. M., ed., *Calendar of Inquisitions Miscellaneous*, 8 vols. (London: H. M. S. O., 1916–2003)

Malden, Henry Elliot, ed., *The Cely Papers: Selections from the Correspondence and Memoranda of the Cely Family, Merchants of the Staple, A.D. 1475–1488* (London: Longmans, Green, and Co., 1900)

Merson, A. L., ed., *The Third Book of Remembrance of Southampton, 1514–1602, Volume 1* (Southampton: University of Southampton, 1952)

Myers, A. R., ed., *English Historical Documents, 1327–1485, Volume 4* (London: Eyre & Spottiswoode, 1969)

Nicolas, Harris, ed., *Proceedings and Ordinances of the Privy Council of England*, 7 vols. (London: Record Commission, 1835)

Oppenheim, M., ed., *Naval Accounts and Inventories of the Reign of Henry VII, 1485–8 and 1495–7* (London: Naval Records Society, 1896)

Preest, David, and James G. Clark, eds., *The Chronica Maiora of Thomas Walsingham 1376–1422* (Woodbridge: Boydell Press, 2005)

Reddaway, T. F., and Alwyn A. Ruddock, eds., 'The Accounts of John Balsall, Purser of the Trinity of Bristol, 1480–1481', *Camden Miscellany*, XXIII. Camden 4th ser. VII (London: Offices of the Royal Historical Society, 1969)

Riley, H. T., ed., *Registra Quorundam Abbatum Monasterii S. Albani, Qui Sæculo XVmo. Floruere. Volume 1, Registrum Abbatiae Johannis Whethamstede, Abbatis Monasterii Sancti Albani, Iterum Susceptae; Robert Blakeney, Capellano, Quondam Adscriptum* (London: Longman & Co., 1872)

Riley, H. T., ed., *Thomae Walsingham, Quondam Monachi S. Albani, Historia Anglicana. Volume 2, A. D. 1381–1422* (Cambridge: Cambridge University Press, 2012)

Rose, Susan, ed., *The Navy of the Lancastrian Kings: Accounts and Inventories of William Soper, Keeper of the King's Ships, 1422–1427* (London: Allen & Unwin for the Navy Record Society, 1982)

Sixth Report of the Royal Commission on Historical Manuscripts (London: H. M. S. O., 1879)

Smith, L. T., ed., *The Maire of Bristowe is Kalendar, by Robert Ricart, Town Clerk of Bristol 18 Edward IV [1479]* (London: Camden Society, 1872)

Smith, L. T., ed., *The Itinerary of John Leland in or about the Years 1535–1543. Parts I to III* (London: George Bell and Sons, 1907)

Stevenson, Joseph, ed., *Letters and Papers Illustrative of the Wars of the English in France during the Reign of Henry the Sixth, King of England*, 3 vols. (London: Longman, Green, Longman, and Roberts, 1861–4)

Stevenson, Joseph, ed., *Narratives of the Expulsion of the English from Normandy, M. CCCC. XLIX.–M.CCC.L* (London: Longman, Green, Longman, Roberts, and Green, 1863)

Taylor, Frank, and John. S. Roskell, eds., *Gesta Henrici Quinti* (Oxford: Oxford University Press, 2004)

Veale, E. W. W., ed., *The Great Red Book of Bristol*, 5 vols. (Bristol: Bristol Record Society, 1931–53)

Worth, R. N., ed., *Calendar of the Plymouth Municipal Records* (Plymouth: William Brendon and Son, 1893)

SECONDARY SOURCES

Allmand, C. T., 'L'Artillerie de L'armée Anglaise et Son Organisation à L'époque de Jeanne d'Arc', in *Jeanne d'Arc, une Époque, un Rayonnement*, ed. R. Pernaud (Paris: CNRS, 1982), pp. 73–83

Allmand, C. T., *Lancastrian Normandy 1415–1450: The History of a Medieval Occupation* (Oxford: Clarendon Press, 1983)

Allmand, C. T., *The Hundred Years War: England and France at War, c.1300–c.1450* (Cambridge: Cambridge University Press, 1988)

Allmand, C. T., *Henry V* (New Haven: Yale University Press, 1997)

Arthurson, Ian, 'The King's Voyage into Scotland: The War that Never Was', in *England in the Fifteenth Century: Proceedings of the 1986 Harlaxton Symposium*, ed. Daniel Williams (Woodbridge: Boydell Press, 1987), pp. 1–22

Arthurson, Ian, *The Perkin Warbeck Conspiracy, 1491–1499* (Stroud: Sutton, 1997)

Attreed, Lorraine, *The King's Towns: Identity and Survival in Late Medieval English Boroughs* (New York: Peter Lang Publishing, 2001)

Awty, Brian, and Christopher Whittick, 'The Lordship of Canterbury, Iron-Founding at Buxted, and the Continental Antecedents of Cannon-Founding in the Weald', *Sussex Archaeological Collections*, 140 (2002), pp. 71–81

Badham, Sally, and John Blatchly, 'The Bellfounder's Indent at Bury St. Edmunds', *Proceedings of the Suffolk Institute of Archaeology and History*, 36 (1988), pp. 288–97

Barron, Caroline M., *London in the Later Middle Ages: Government and People, 1200–1500* (Oxford: Oxford University Press, 2004)

Bennett, M. J., 'Henry VII and the Northern Rising of 1489', *The English Historical Review*, 105 (1990), pp. 34–59

Blackmore, Howard L., *The Armouries of the Tower of London, I: Ordnance* (London: H. M. S. O., 1976)

Blair, Claude, and John Blair, 'Copper Alloys', in *English Medieval Industries*, ed. John Blair and N. Ramsay (London: Hambledon Press, 1991), pp. 81–106

Boffa, Sergio, *Warfare in Medieval Brabant, 1356–1406* (Woodbridge: Boydell Press, 2004)

Bradbury, J., *The Medieval Siege* (Woodbridge: Boydell Press, 1992)

Brackenbury, H., 'Ancient Cannon in Europe. Part I: From their First Employment to A.D. 1350', *Proceedings of the Royal Artillery Institution*, 4 (1865), pp. 1–22

Brackenbury, H., 'Ancient Cannon in Europe. Part II: From A.D. 1351 to 1400', *Proceedings of the Royal Artillery Institution*, 5 (1867), pp. 23–59

Bree, John, *The Cursory Sketch of the State of the Naval, Military, and Civil Establishment, Legislative, Judicial, and Domestic Economy of this Kingdom during the Fourteenth Century;*

with a Particular Account of the Campaign of King Edward the Third, in Normandy and France, in the Years 1345 and 1346 (London: Printed for the Author, 1791)

Brown, A. L., *The Governance of Late Medieval England 1272–1461* (London: Edward Arnold, 1989)

Brown, R. A., *English Castles* (Woodbridge: Boydell Press, 2004)

Carpenter-Turner, W. J., 'The Building of the Gracedieu, Valentine and Falconer at Southampton, 1416–1420', *The Mariner's Mirror*, 40 (1954), pp. 55–72

Chrimes, S. B., 'Some Letters of John of Lancaster as Warden of the East Marches Towards Scotland', *Speculum*, 14 (1939), pp. 3–27

Chrimes, S. B., *Henry VII* (New Haven and London: Yale University Press, 1999)

Christiansen, Eric, *The Northern Crusades* (London: Penguin Books, 1997)

Clarke, Helen, *Sandwich: The 'Completest Medieval Town in England': A Study of the Town and Port from its Origins to 1600* (Oxford: Oxbow, 2010)

Colvin, H. M., *The History of the King's Works*. 6 vols. (London: H. M. S. O., 1963–82)

Contamine, Philippe, 'L'Artillerie Royale Francaise á la Veille des Guerres d'Italie', *Annales de Bretagne*, 71 (1964), pp. 221–61

Contamine, Philippe, *War in the Middle Ages*, trans. Michael Jones (Oxford: Blackwell, 1984)

Cook, Weston F., 'The Cannon Conquest of Nasrid Spain and the End of the Reconquista', *The Journal of Military History*, 57 (1993), pp. 43–70

Crawford, Anne, *Yorkist Lord: John Howard, Duke of Norfolk, c.1425–1485* (London: Continuum, 2010)

Creighton, O. H., *Medieval Towns Walls: An Archaeology and Social History of Urban Defence* (Stroud: Tempus, 2005)

Cressy, David, *Saltpeter: The Mother of Gunpowder* (Oxford: Oxford University Press, 2013)

Crouy-Chanel, Emmanuel de, *Canons Médiévaux: Puissance du Feu* (Paris: Rempart, 2010)

Cunningham, Sean, 'National War and Dynastic Politics: Henry VII's Capacity to Wage War in the Scottish Campaigns of 1496–7', in *England and Scotland at War, c.1296–c.1513*, ed. Andy King and David Simpkin (Leiden: Brill, 2012), pp. 297–328

Currin, J. M., '"The King's Army into the Partes of Bretaigne": Henry VII and the Breton Wars, 1489–1491', *War in History*, 7 (2000), pp. 379–412

Curry, Anne, 'English Armies in the Fifteenth Century', in *Arms, Armies and Fortifications in the Hundred Years War*, ed. Anne Curry and Michael Hughes (Woodbridge: Boydell Press, 1994), pp. 39–68

Curry, Anne, *The Hundred Years' War: 1337–1453* (Oxford: Osprey, 2002)

Curry, Anne, 'The "Coronation Expedition" and Henry VI's Court in France, 1430 to 1432', in *The Lancastrian Court*, ed. Jenny Stratford (Donington: Shaun Tyas, 2003), pp. 29–52

Curry, Anne, *Agincourt: A New History* (Stroud: Tempus, 2005)

Curry, Anne, 'Guns and Goddams: Was there a Military Revolution in Lancastrian Normandy, 1415–50?', *Journal of Medieval Military History*, 8 (2010), pp. 171–88

Curry, Anne, 'New Regime, New Army? Henry IV's Scottish Expedition of 1400', *The English Historical Review*, 125 (2010), pp. 1382–1413

Cushway, Graham, *Edward III and the War at Sea: The English Navy, 1327–1377* (Woodbridge: Boydell Press, 2011)

Davies, C. S. L., 'Richard III, Brittany, and Henry Tudor 1483–1485', *Nottingham Medieval Studies*, 37 (1993), pp. 110–26

Davies, R. R., *The Revolt of Owain Glyn Dŵr* (Oxford: Oxford University Press, 1995)

Davies, Silvester, J., *A History of Southampton: Partly from the MS. Of Dr. Speed in the Southampton Archives* (Southampton: Gilbert, 1883)

DeVries, Kelly, *Medieval Military Technology* (Peterborough, Ontario: Broadview Press, 1992)

DeVries, Kelly, 'The Impact of Gunpowder Weaponry on Siege Warfare in the Hundred Years War', in *The Medieval City under Siege*, ed. Ivy A. Corfis and Michael Wolfe (Woodbridge: Boydell Press, 1995), pp. 227–44

DeVries, Kelly, 'The Technology of Gunpowder Weaponry in Western Europe During the Hundred Years War', in *XXII. Kongreß Der Internationalen Kommission Für Militärgeschichte Acta 22: Von Crécy Bis Ohács Kriegswesen Im Späten Mittelalter (1346–1526)* (Vienna: Heeresgeschichtliches Museum, 1997), pp. 285–98

DeVries, Kelly, 'The Effectiveness of Fifteenth-Century Shipboard Artillery', *The Mariner's Mirror*, 84 (1998), pp. 389–99

DeVries, Kelly, 'Gunpowder Weaponry and the Rise of the Early Modern State', *War in History*, 5 (1998), pp. 127–45

DeVries, Kelly, 'The Use of Gunpowder Weapons in the Wars of the Roses', in *Traditions and Transformations in Late Medieval England*, ed. Douglas Biggs (Leiden: Brill, 2001), pp. 21–38

DeVries, Kelly, 'Facing the New Technology: Gunpowder Defenses in Military Architecture Before the Trace Italienne, 1350–1500', in *The Heirs of Archimedes: Science and the Art of War through the Age of Enlightenment*, ed. Brenda J. Buchanan (London: Massachusetts Institute of Technology, 2005), pp. 37–72

Foard, Glenn, and Anne Curry, *Bosworth 1485: A Battlefield Rediscovered* (Oxford: Oxbow Books, 2013)

Foard, Glenn, and Richard Morris, *The Archaeology of English Battlefields: Conflict in the Pre-Industrial Landscape* (York: Council for British Archaeology, 2012)

Friel, Ian, *The Good Ship: Ships, Shipbuilding and Technology in England, 1200–1520* (London: British Museum Press, 1995)

Friel, Ian, 'Oars, Sails and Guns: The English and War at Sea, c.1200–c.1500', in *War at Sea in the Middle Ages and Renaissance*, ed. John B. Hattendorf and Richard W. Unger (Woodbridge: Boydell & Brewer, 2003), pp. 69–80

Geddes, Jane, 'Iron', in *English Medieval Industries*, ed. John Blair and N. Ramsay (London: Hambledon Press, 1991), pp. 167–88

Given-Wilson, Chris, *Henry IV* (New Haven: Yale University Press, 2016)

Goodman, Anthony, *The Wars of the Roses: Military Activity and English Society, 1452–97* (London: Routledge, 1981)

Griffiths, Ralph A., *The Reign of King Henry VI: The Exercise of Royal Authority, 1422–1461* (London: Benn, 1981)

Grummitt, David, 'The Defence of Calais and the Development of Gunpowder Weaponry in England in the Late Fifteenth Century', *War in History*, 7 (2000), pp. 253–72

Grummitt, David, *The Calais Garrison: War and Military Service in England, 1436–1558* (Woodbridge: Boydell Press, 2008)

Grummitt, David, 'A Military Revolution in the North? The Impact of Gunpowder Weaponry on the Anglo-Scottish Marches', in *England and Scotland at War, c. 1296–c. 1513*, ed. Andy King and David Simpkin (Leiden-Boston: Brill Academic Publishers, 2012), pp. 283–96

Gunn, Steven, David Grummitt and Hans Cools, *War, State and Society in England and the Netherlands, 1477–1559* (Oxford: Oxford University Press, 2006)

Hall, Bert S., *Weapons and Warfare in Renaissance Europe* (London: The John Hopkins Press Ltd, 1997)

Hogg, O. F. G., *English Artillery 1326–1716* (London: Royal Artillery Institution, 1963)

Hogg, O. F. G., *Artillery: Its Origin, Heyday and Decline* (London: C. Hurst, 1970)

Howlett, Richard, 'Norwich Artillery in the Fourteenth Century', *Norfolk Archaeology*, XVI (1907), pp. 46–75

Hunter, Joseph, 'Proofs of the Early Use of Gunpowder in the English Army', *Archaeologia*, 32/2 (1846), pp. 379–87

Hutchinson, Michael, 'Edward IV's Bulwark: Excavations at Tower Hill, London, 1985', *Transactions of the London and Middlesex Archaeological Society*, 47 (1996), pp. 103–44

Jones, Michael K, 'John Beaufort, Duke of Somerset, and the French Expedition of 1443', in *Patronage, the Crown and the Provinces*, ed. R. A. Griffiths (Gloucester: Sutton, 1981), pp. 79–102

Keen, Maurice H., 'The Changing Scene: Guns, Gunpowder, and Permanent Armies', in *Medieval Warfare: A History*, ed. Maurice Keen (Oxford: Oxford University Press, 1999), pp. 273–92

Kenyon, John R., 'Early Artillery Fortifications in England and Wales: A Preliminary Survey and Reappraisal', *Archaeological Journal*, 138 (1981), pp. 205–40

King, Andy, 'Gunners, Aides and Archers: The Personnel of the English Ordnance Companies in Normandy in the Fifteenth Century', *Journal of Medieval Military History*, 9 (2011), pp. 65–75

Kirby, J. L., *Henry IV of England* (London: Constable, 1970)

Kleineke, Hannes, 'Gerhard von Wesel's Newsletter from England, 17 April 1471', *The Ricardian*, 16 (2006), pp. 66–83

Kowaleski, Maryanne, 'Warfare, Shipping, and Crown Patronage: The Impact of the Hundred Years War on the Port Towns of Medieval England', in *Money, Markets and Trade in Late Medieval Europe: Essays in Honour of John H. A. Munro*, ed. Lawrin Armstong, Ivana Elbl and Martin M. Elbl (Leiden and Boston: Brill, 2007), pp. 233–56

Latham, R. E., *Revised Medieval Latin Word-List: From British and Irish Sources* (London: Oxford University Press, 1965)

Laughton, Jane, *Life in a Late Medieval City* (Oxford: Windgather Press, 2008)

Mallett, M. E., and J. R. Hale, *The Military Organization of a Renaissance State: Venice c. 1400 to 1617* (Cambridge: Cambridge University Press, 1984)

Mayhew, G. J., 'Rye and the Defence of the Narrow Seas: A 16th-Century Town at War', *Sussex Archaeological Collections*, 122 (1984), pp. 107–26

McCarthy, Michael R., *Carlisle Castle: A Survey and Documentary History* (London: English Heritage, 1990)

Moffett, Randall, 'Military Equipment in the Town of Southampton during the Fourteenth and Fifteenth Centuries', *Journal of Medieval Military History*, 9 (2011), pp. 167–99

Moore, Alan, 'Accounts and Inventories of John Starlyng, Clerk of the King's Ships to Henry IV', *Mariner's Mirror*, 4 (1914), pp. 20–6

Murray, K. M. Elisabeth, *The Constitutional History of the Cinque Ports* (Manchester: Manchester University Press, 1935)

Needham, Joseph, Ho Ping Yú, Lu Gwei-Djen and Wang Ling, *Science and Civilisation in China. Vol. 5, Chemistry and Chemical Technology, part 7, Military Technology: The Gunpowder Epic* (Cambridge: Cambridge University Press, 1986)

Newhall, Richard, *The English Conquest of Normandy 1416–1424: A Study in Fifteenth Century Warfare* (New Haven: Yale University Press, 1924)

Norman, A. V. B., 'Notes on Some Early Representations of Guns and on Ribaudekins', *Journal of the Arms and Armour Society*, 8 (1975), pp. 234–7

Oman, Charles, *A History of the Art of War in the Middle Ages, Volume Two: 1278–1485 AD* (London: Greenhill Books, 1998)

O'Neil, B. H. St J., *Castles and Cannon: A Study of Early Artillery Fortifications in England* (Oxford: Clarendon Press, 1960)

Onions, C. T., ed., *The Shorter Oxford English Dictionary*, 2 vols. (Oxford: Clarendon Press, 1973)

Ormrod, W. M., 'The Protecolla Rolls and English Government Finance, 1353–1364', *The English Historical Review*, 102 (1987), pp. 622–32

Parker, Geoffrey, *The Military Revolution: Military Innovation and the Rise of the West, 1500–1800* (Cambridge: Cambridge University Press, 1988)

Pépin, Guilhem, 'The French Offensives of 1404–1407 against Anglo-Gascon Aquitaine', *Journal of Medieval Military History*, 9 (2011), pp. 1–40

Pepper, Simon, 'Aspects of Operational Art: Communications, Cannon, and Small War', in *European Warfare, 1350–1750*, ed. Frank Tallett and D. J. B. Trim (Cambridge: Cambridge University Press, 2010), pp. 181–202

Peter, Richard, and Otho Bathurst Peter, *The Histories of Launceston and Dunheved, in the County of Cornwall* (Plymouth: Brendon & Son, 1885)

Petrovic, Djurdjica, 'Fire-arms in the Balkans on the Eve of and After the Ottoman Conquests of the Fourteenth and Fifteenth Centuries', in *War, Technology and Society in the Middle East*, ed. V. J. Parry and M. E. Yapp (London: Oxford University Press, 1975), pp. 164–94

Platt, Colin, *Medieval Southampton: The Port and Trading Community, A.D. 1000–1600* (London: Routledge and Kegan Paul, 1973)

Pollard, Hugh, *Pollard's History of Firearms* (London: Country Life Books, 1983)

Pounds, N. J. G., *The Medieval Castle in England and Wales: A Social and Political History* (Cambridge: Cambridge University Press, 1990)

Prestwich, Michael, *Armies and Warfare in the Middle Ages: The English Experience* (New Haven and London: Yale University Press, 1996)

Prestwich, Michael, 'The Battle of Crécy', in *The Battle of Crécy, 1346*, ed. Andrew Ayton and Philip Preston (Woodbridge: Boydell and Brewer, 2005), pp. 139–57

Richardson, Thom, *The Tower Armoury in the Fourteenth Century* (Leeds: Royal Armouries, 2016)

Richmond, C. F., 'The Keeping of the Seas during the Hundred Years War: 1422–1440', *History*, 49 (1964), pp. 283–98

Richmond, C. F., 'English Naval Power in the Fifteenth Century', *History*, 52 (1967), pp. 1–15
Richmond, C. F., 'Fauconberg's Kentish Rising of May 1471', *The English Historical Review*, 85 (1970), pp. 673–92
Rimer, Graeme, 'Early Handguns', *Royal Armouries Yearbook*, 1 (1996), pp. 73–8
Roberts, Michael, *Essays in Swedish History* (London: Weidenfeld & Nicolson, 1967)
Rogers, Clifford J., 'The Military Revolutions of the Hundred Years' War', *Journal of Military History*, 57 (1993), pp. 241–78
Rogers, James E. Thorold, *A History of Agriculture and Prices in England from the Year after the Oxford Parliament, (1259) to the Commencement of the Continental War (1793)*, 8 vols. (Oxford: Clarendon Press, 1866–1902)
Ross, Charles, *Edward IV* (London: Eyre Methuen, 1974)
Ross, Charles, *Richard III* (London: Eyre Methuen, 1981)
Ruddock, Alwyn A., *Italian Merchants and Shipping in Southampton, 1270–1600* (Southampton: University College, 1951)
Saul, Nigel, *Richard II* (New Haven and London: Yale University Press, 1997)
Saunders, A., 'The Cow Tower, Norwich: An East Anglican Bastille?', *Medieval Archaeology*, XXIX (1985), pp. 109–19
Scofield, Cora, L., *The Life and Reign of Edward the Fourth: King of England and of France and Lord of Ireland*, 2 vols. (London: Longmans, Green, 1923)
Sherborne, James, 'The Hundred Years' War. The English Navy: Shipping and Manpower 1369–1489', *Past and Present*, 37 (1967), pp. 163–75
Sherborne James, 'The Defence of the Realm and the Impeachment of Michael de la Pole in 1386', in *Politics and Crisis in Fourteenth-Century England*, ed. John Taylor and Wendy Childs (Gloucester: Alan Sutton, 1990), pp. 99–118
Sicking, Louis, 'Naval Warfare in Europe, c. 1330–1680', in *European Warfare, 1350–1750*, ed. Frank Tallett and D. J. B. Trim (Cambridge: Cambridge University Press, 2010), pp. 236–63
Smith, Robert D., 'The Technology of Wrought-Iron Artillery', *Royal Armouries Yearbook*, 5 (2000), pp. 68–79
Smith, Robert D., '"All Manner of Peeces" Artillery in the Late Medieval Period', *Royal Armouries Yearbook*, 7 (2002), pp. 130–8
Smith, Robert D., and Kelly DeVries, *The Artillery of the Dukes of Burgundy 1363–1477* (Woodbridge: Boydell Press, 2005)
Spencer, Dan, 'The Lancastrian Armament Programme of the 1450s and the Development of Field Guns', *The Ricardian*, 25 (2015), pp. 61–70
Spencer, Dan, 'The Provision of Artillery for the 1428 Expedition to France', *Journal of Medieval Military History*, 13 (2015), pp. 179–92
Spencer, Dan, 'The Tower of London and Firearms in the Reign of Edward IV', *Arms & Armour*, 13:2 (2016), pp. 98–110
Spencer, Dan, 'Royal Castles and Coastal Defence in the Late Fourteenth Century', *Nottingham Medieval Studies*, 61 (2017), pp. 147–70
Stoyle, Mark, *Circled with Stone: Exeter's City Walls, 1485–1660* (Exeter: University of Exeter Press, 2003)
Sumption, Jonathan, *The Hundred Years War: Volume I, Trial by Battle* (London: Faber, 1990)

Sumption, Jonathan, *The Hundred Years War: Volume III, Divided Houses* (London: Faber, 2009)

Sumption, Jonathan, *The Hundred Years War: Volume IV, Cursed Kings* (London: Faber, 2015)

Thompson, M. W., *The Decline of the Castle* (Cambridge: Cambridge University Press, 1987)

Tout, T. F., 'Firearms in England in the Fourteenth Century', *The English Historical Review*, 26 (1911), pp. 666–702

Tout, T. F., *Chapters in the Administrative History of Medieval England*, 6 vols. (Manchester: The University Press, 1920–33)

Tuck, Anthony, 'Henry IV and Chivalry', in *Henry IV: The Establishment of the Regime, 1399–1406*, ed. Gwilym Dodd and Douglas Biggs (York: York Medieval Press, 2003), pp. 55–72

Turner, Hilary L., *Town Defences in England and Wales: An Architectural and Documentary Study AD 900–1500* (London: John Baker Ltd, 1971)

Vale, M. G. A., *War and Chivalry: Warfare and Aristocratic Culture in England, France and Burgundy at the End of the Middle Ages* (London: Duckworth, 1981)

Vaughan, Richard, *John the Fearless: The Growth of Burgundian Power* (London: Longmans, 1966)

Vaughan, Richard, *Philip the Good: The Apogee of Burgundy* (London: Longmans, Green and Co Ltd, 1970)

Vaughan, Richard, *Charles the Bold: The Last Valois Duke of Burgundy* (Woodbridge: Boydell Press, 2002)

Veszprémy, László, 'The Birth of Military Science in Hungary: The Period of the Anjou and Luxembourg Kings', in *A Millennium of Hungarian Military History*, ed. László Veszprémy and Béla K. Király (New York: Columbia University Press, 2002), pp. 26–53

Weinbaum, Martin, *The Incorporation of Boroughs* (Manchester: Manchester University Press, 1937)

Welch, Charles, *History of the Tower Bridge* (London: Smith, Elder and Co., 1894)

Whelan, Mark, 'Catastrophe or Consolidation? Sigismund's Response to Defeat After the Crusade of Nicopolis', in *Between Worlds: The Age of the Jagiellonians*, ed. Florian Ardelean, Christopher Nicholson and Johannes Preiser-Kapeller (Frankfurt am Main: Peter Lang, 2013), pp. 215–27

Wylie, James *Hamilton, History of England Under Henry the Fourth. Vol. I, 1399–1404* (London: Longmans, Green and Co., 1884)

Wylie, James Hamilton, *History of England Under Henry the Fourth. Vol. II, 1405–1406* (London: Longmans, Green and Co, 1894)

Wylie, James Hamilton, *History of England Under Henry the Fourth. Vol. III, 1407–1410* (London: Longmans, Green and Co, 1896)

Wylie, James Hamilton, *History of England Under Henry the Fourth. Vol. IV, 1411–1413* (London: Longmans, Green and Co, 1898)

Wylie, James Hamilton, *The Reign of Henry the Fifth. Vol. III, 1415–1422* (Cambridge: Cambridge University Press, 1929)

UNPUBLISHED THESES

Caldwell, D. H., 'Guns in Scotland: The Manufacture and Use of Guns and their Influence on Warfare from the Fourteenth Century to c. 1625' (unpublished doctoral thesis, University of Edinburgh, 1982)

Cushway, Graham Russell, '"The Lord of the Sea": The English Navy in the Reign of Edward III' (unpublished doctoral thesis, University of Exeter, 2006)

Dimmock, Spencer, 'Class and the Social Transformation of a Late Medieval Small Town: Lydd c. 1450–1550' (unpublished doctoral thesis, University of Kent, 1998)

Homer, Paul Leroy, 'Studies in the Military Organization of the Yorkist Kings' (unpublished doctoral thesis, University of Minnesota, 1977)

Jones, M. K., 'The Beaufort Family and the War in France, 1421–1450' (unpublished doctoral thesis, University of Bristol, 1982)

Moffett, Randall, 'The Military Organization of Southampton in the Late Medieval Period, 1300–1500' (unpublished doctoral thesis, University of Southampton, 2009)

Rainey Jr, John Riley, 'The Defence of Calais, 1436–1477' (unpublished doctoral thesis, Rutgers University, 1987)

Spencer, Dan, 'The Development of Gunpowder Weapons in Late Medieval England' (unpublished doctoral thesis, University of Southampton, 2016)

Walton, Steven A., 'The Art of Gunnery in Renaissance England' (unpublished doctoral thesis, University of Toronto, 1999)

Warner, Mark, William, 'The Montagu Earls of Salisbury, 1300–1428' (unpublished doctoral thesis, University College London, 1991)

Winston, James E., 'English Towns in the Wars of the Roses' (unpublished doctoral thesis, University of Pennsylvania, 1921)

ELECTRONIC RESOURCES

AHRC-funded database, *The Soldier in Later Medieval England*, http://www.icmacentre.ac.uk/soldier/database/ [accessed 23 August 2014]

Brewer, J. S., ed., 'Henry VIII: December 1509', in *Letters and Papers, Foreign and Domestic, Henry VIII, Volume 1, 1509–1514* (London, 1920), pp. 127–44, http://www.british-history.ac.uk/letters-papers-hen8/vol1/pp127-144 [accessed 29 January 2015]

Cunningham, Sean, 'Guildford, Sir Richard (c.1450–1506)', *Oxford Dictionary of National Biography, Oxford University Press*, 2004; online edn, Jan. 2008, http://www.oxforddnb.com/view/article/11723 [accessed 1 April 2015]

Curry, Anne, 'Kyriell, Sir Thomas (1396–1461)', *Oxford Dictionary of National Biography, Oxford University Press*, 2004; online edition, Jan. 2008, http://www.oxforddnb.com/view/article/50135?docPos=1 [accessed 3 January 2013]

Gunn, Steven, 'Vere, John de, Thirteenth Earl of Oxford', *Oxford Dictionary of National Biography, Oxford University Press, 2004*; online edition, Jan. 2008, http://www.oxforddnb.com/view/article/28214?docPos=3 [accessed 22 December 2014]

Harriss, G. L., 'Beaufort, Thomas, Duke of Exeter (1377?–1426)', in *Oxford Dictionary of National Biography*, http://www.oxforddnb.com/view/article/1864?&docPos=30&backToResults=/search/refine/?docStart=1 [accessed 25 August 2014]

Rawcliffe, Carole, 'Blount, Sir Walter (d. 1403)', *Oxford Dictionary of National Biography, Oxford University Press, 2004*, http://www.oxforddnb.com/view/article/2699 [accessed 7 April 2015]

Records of the Duchy of Lancaster, http://discovery.nationalarchives.gov.uk/details/r/C83 [accessed 11 April 2018]

Records of Various Departments, Arranged Artificially According to Type, and Formerly Entitled Special Collections, http://discovery.nationalarchives.gov.uk/details/r/C246 [accessed 11 April 2018]

Index